THE CONSTANTS OF PHYSICS

THE CONSTANTS OF PHYSICS

PROCEEDINGS OF
A ROYAL SOCIETY DISCUSSION MEETING
HELD ON 25 AND 26 MAY 1983

ORGANIZED BY W. H. McCREA, F.R.S.,
M. J. REES, F.R.S., AND S. WEINBERG, FOR.MEM.R.S.,
AND EDITED BY W. H. McCREA, F.R.S., AND
M. J. REES, F.R.S.

LONDON
THE ROYAL SOCIETY
1983

Printed in Great Britain for the Royal Society
at the
University Press, Cambridge

ISBN 0 85403 224 X

First published in *Philosophical Transactions of the Royal Society of London*,
series A, volume 310 (no. 1512), pages 209–363

Copyright

© 1983 The Royal Society and the authors of individual papers.

It is the policy of the Royal Society not to charge any royalty for the production of a single copy of any one article made for private study or research. Requests for the copying or reprinting of any article for any other purpose should be sent to the Royal Society.

Published by the Royal Society
6 Carlton House Terrace, London SW1Y 5AG

PREFACE

The present time may well prove to be the greatest epoch hitherto as regards delving into the foundations of physical science. Such delving acquires definite purpose and precision because of the existence of the *constants* of physics. Physicists seek to understand these constants at various levels of sophistication. Are there discoverable relations between the constants at one level revealed by relating these to constants at a more fundamental level? Is there a level at which no constants are needed, so that the Universe can be inferred to be as it is simply because it exists? But how does a changing Universe give rise to unchanging constants; or does it? Again, by looking at the Universe from the standpoint of everyday physics, it is seen that it would be a vastly different place, were the constants to have values only a little different from the values measured in the laboratory. Is this another indication that the constants must have these values in order for the Universe to exist? Or have we to entertain the notion of the 'existence' – in some sense – of infinitely many other universes? All such enquiries might seem to be without much meaning were it not for the most remarkable development proceeding at the present time wherein the most profound discoveries in particle physics appear to fit perfectly with the concept of the very early Universe to which physicists and astronomers are led by the latest discoveries about the astronomical Universe in the large.

Such considerations made it appear timely to propose holding a Meeting for Discussion of these topics and to seek to make the report of its proceedings available as soon as poosible. The object was to try to bring together the chief lines of current thinking in the field, rather than mainly to collate accepted results of experiments or theories.

The Editors believe that in the form in which the authors have presented their contributions they collectively convey the spirit of the meeting itself. They express their appreciation of the authors' cooperation in the effort to achieve early publication.

In preparing for the meeting the Editors as organizers were joined by Professor Steven Weinberg, For.Mem.RS.; they record their indebtedness to him for his wise advice before and during the meeting. They are grateful to a number of other colleagues whose advice they sought, and very particularly to Sir Denys Wilkinson, F.R.S., whose thoughtful interest throughout was invaluable.

The organizers record their thanks to the Society's staff for their cheerful help in all the preparations, particularly to Miss Christine Johnson for her skill in organizing the meeting They thank Dr Marion Storm who has seen the volume through the press so expertly and so expeditiously, and the Printers who have cooperated with her with so much professional perfection.

October 1983

W. H. McCREA
M. J. REES

CONTENTS

	PAGE
PREFACE	[v]

W. H. McCREA, F.R.S.
    Introductory remarks   [1]

K. F. SMITH
    The measurement of the fundamental constants   [5]

B. W. PETLEY
    Towards the next evaluation of the fundamental physical constants   [11]

M. GOLDHABER
    The search for proton decay and other rare phenomena   [15]

R. D. REASENBERG
    The constancy of G and other gravitational experiments   [17]

J. M. IRVINE
    Limits on the variability of coupling constants from the Oklo natural reactor   [29]

B. E. J. PAGEL
    Implications of quasar spectroscopy for constancy of constants   [35]

S. WEINBERG, For.Mem.R.S.
    Overview of theoretical prospects for understanding the values of fundamental constants   [39]
    Discussion: H. B. NIELSEN, J. G. TAYLOR   [41]

C. H. LLEWELLYN SMITH
    The strong, electromagnetic and weak couplings   [43]
    Discussion: H. B. NIELSEN   [49]

H. B. NIELSEN
    Field theories without fundamental gauge symmetries   [51]
    Discussion: J. G. TAYLOR   [61]

S. L. ADLER
    Einstein gravitation as a long wavelength effective field theory   [63]

J. ELLIS
    Unification and supersymmetry   [69]

CONTENTS

T. W. B. KIBBLE, F.R.S. — PAGE
    Phase transitions in the early Universe and their consequences — [83]
    Discussion: S. KASDAN — [91]

S. W. HAWKING, F.R.S.
    The cosmological constant — [93]
    Discussion: H. B. NIELSEN — [99]

M. J. REES, F.R.S.
    Large numbers and ratios in astrophysics and cosmology — [101]
    Discussion: W. H. MCCREA, F.R.S. — [111]

W. H. PRESS AND A. P. LIGHTMAN
    Dependence of macrophysical phenomena on the values of the fundamental constants — [113]
    Discussion: SIR RUDOLPH PEIERLS, F.R.S., T. GOLD, F.R.S. — [124]

J. D. BARROW
    Dimensionality — [127]

B. CARTER, F.R.S.
    The anthropic principle and its implications for biological evolution — [137]
    Discussion: W. H. MCCREA, F.R.S. — [153]

Introductory remarks

By W. H. McCrea, F.R.S.

Astronomy Centre, University of Sussex, Brighton BN1 9QH, U.K.

What distinguishes modern physics from classical physics is the recognition of the role of fundamental (or universal) constants. Mathematical physics must be formulated so as to admit such constants; that is what distinguishes it from other applied mathematics. It is the particular values actually possessed by the constants that make our Universe what it is. Some analysis of this whole situation is the theme of this Discussion.

We contemplate essentially dimensionless constants, or, equivalently, constants expressed in natural units which exist because the constants exist. Naturally, however, values expressed in 'practical' units are an indispensable convenience.

The domain is one in which observation and theory are inseparable. For instance, had general relativity come without newtonian theory having been thought of, we should not have heard of the gravitational constant G. In this Discussion we learn about observations designed to test whether G varies with time. Now exactly the same observational procedures could be performed by astronomers who had never heard of G. They would express the purpose of the observations in other language. But this language would depend again on whether they had heard of cosmic time or not. Actually, however, in practice a different theoretical approach would probably have led to somewhat differently designed observations. Anyhow, the contemplation of such an example serves to illustrate how theory and observation interact.

At any point in our deliberation, it therefore seems inevitable that we should speak in terms of some definite *theoretical model* of the world of experience. There appears, however, to be no meaning in supposing there to exist a unique final model that we are trying to *discover*. We *construct* a model, we do not discover it.

Is it nevertheless true to say that models lead to the discovery of constants of physics? The answer seems to be yes, in the sense that a model may suggest a set of operations that is found to lead to the 'determination' of some constant. If so, then ever more refined repeatable observations are found to lead to a more and more 'accurate' value.

For, at any rate, most of our present purposes, I think it will be found that a constant of physics has this *operational* meaning. Any such constant is then entered in a table in decimal notation as a value with a standard error, and, of course, the specification of the physical units. Within our present understanding of physics, I would suggest that this is all that a 'constant' can ever mean. We cannot believe in the existence of a celestial 'Landolt–Börnstein' table in which every entry is a mathematically exactly defined number.

Having thus tried to say what is meant by the constants of physics, important general questions spring to mind that we wish to ask about them. Some of these prompted the proposal to hold the present meeting, and resulted in the sequence of topics on the programme. It may be useful briefly to indicate the pattern – a tolerably coherent pattern, as the organizers hope – as follows. The numbers are those of the contributions (here identified also by authors' names) as they were listed in the programme; I hope the contributors will regard my short descriptions as valid so far as they go.

Observation

Laboratory: values of the constants

1. Principles guiding experimental determination and accuracy sought therein. We learn about the nature of recent remarkable advances, although a systematic survey of the results is outside the scope of the discussion (Smith). Experimental data likely to be considered in next evaluation of the constants (Petley).

2. Constants concerned in rare phenomena that have acquired crucial special significance in recent times (Goldhaber).

Astronomy and geophysics: constancy of the constants

3. Results of precision observations in solar system astronomy (Reasenberg).

4. Exploitation of results of a naturally occurring geophysical experiment about 2×10^9 years ago (Irvine).

5. Exploitation of astronomical phenomena of some 2×10^{10} years ago (Pagel).

Theory

The constants in physics: field theories

The field theory aspect of physics is paramount in such studies at the present time.

6. Understanding the values. This survey makes precise, among other matters, some of those so briefly touched upon in this Introduction. Also it explores some possibilities for developments in the near future (Weinberg).

7. Possibility of a unified treatment of certain fields yielding relations between the constants: problems outstanding (Llewellyn Smith).

8. Field theories: their nature and genesis (Nielson).

9. Gravitation: a new view of its status as a field theory. A discussion like this is bound to give special attention to gravitation which is both the most familiar and the least understood field phenomenon in physics (Adler).

10. Unified theories: survey of the latest developments and their significance for determining and understanding the fundamental constants (Ellis).

The constants in cosmology

11. Early universe and its possible phase transitions. This is where certain of the concepts under consideration may find their most significant application (Kibble).

12. The problem of the cosmical constant – also mentioned in the preceding contribution – in relation to fundamental symmetries (Hawking).

13*. The origin and significance of certain 'cosmological numbers': possible relation to the constants of physics (Rees).

14*. The dependence upon the values of the constants of physics of macro-phenomena on the Earth and in the cosmos (Press and Lightman).

15. Dependence of physics upon the basic constant of dimensionality (Barrow).

16*. The anthropic principle and the significance for physical and biological theory of the values of the constants of physics (Carter).

Interest in the topics marked* in recent times originally instigated the proposal to hold this

Discussion. However, because of their speculative character and of their inability as yet to produce new predictions, it was considered that the main emphasis ought to be upon the study of the constants themselves rather than the role of the constants in these applications.

One hopes that this sketch shows there to be some logic in the general structure of the programme, even if not always in the precise sequence of the papers. However that may be, we may seem to be tackling almost everything in basic physics and in cosmology all at once. This is not the normally recommended way to achieve solid progress. But the ambition of physicists *is* to accomplish some grand unification; the following pages may serve to show what headway can now be claimed.

The measurement of the fundamental constants

By K. F. Smith

School of Mathematical and Physical Sciences, University of Sussex,
Brighton BN1 9QH, U.K.

Some recent precision experiments that are likely to influence the accepted values of the fundamental constants are reviewed briefly: the measurement of the velocity of light, the possibility of redefining the metre in terms of the caesium time standard, developments that may allow the introduction of an atomic mass standard, the use of the Josephson effect to maintain electrical standards, and some experiments that have led to an improved precision for the fine structure constant.

THE EXPERIMENTAL MEASUREMENT OF FUNDAMENTAL CONSTANTS

Measurement of physical quantities

The measurement of any physical quantity always requires in effect an apparatus capable of comparing what is to be measured with a standard unit of the same quantity. It is important that the standard be reproduceable in different countries, that it should not drift with time or be disturbed by external influences such as temperature. The measuring apparatus acting as a comparator or null detector must have sufficient sensitivity to achieve the precision required; it should be possible to estimate the magnitudes of the known systematic errors associated with the system so that in the end the precision is limited by the random effects introduced by the standard and the measuring apparatus. The standard error quoted is a measure of the overall uncertainty in the sense that a repeat measurement is unlikely to fall outside three standard deviations from the result given.

It is the job of the laboratories such as The National Physical Laboratory in England (N.P.L.) to maintain standards of the highest precision possible so that the measuring apparatus used in the individual laboratories in universities and industry can be calibrated. Resistance and voltage, for example, can be measured to one part in 10^6, time delays of many seconds to about one microsecond and frequency to the order one part in 10^{12}. During the last twenty years there has been a significant increase in the accuracy available for many measurements as direct result of a number of technical developments and the automation possible with modern electronics.

Fundamental constants

Many of the measurements made in the physical sciences are designed to test theories: theories that predict the way some part of the solar system might develop, for example, models of super-conductivity, or the way electromagnetic radiation interacts with atoms in a laser. The theory often predicts relations between directly measurable quantities such as frequency, magnetic field, temperature and the so called 'fundamental constants', the set of quantities that, at any time, combined with the appropriate theory, should be able to explain at both the microscopic and the macroscopic level the behaviour of physical systems. Some of the more obvious fundamental constants are G, the gravitational constant, c the speed of light, h Planck's constant, e the charge on the electron, m_e the mass of the electron, m_p the mass of the proton, and k Boltzmann's constant. Developments in fundamental particle physics have required the introduction of another group of fundamental constants which cannot be predicted in terms of those already mentioned, constants such as G_v the coupling constant for weak nuclear interactions, the masses of various mesons and the Weinberg angle, which relates the strengths of charged and neutral weak currents.

The number of fundamental constants in use tends to increase with time as the experiments and theories become more diverse, but there is always the hope that additional relations between the constants will emerge when new unification theories are developed. The magnetic moment of the proton, which can be measured as accurately as we can reproduce a unit of magnetic field, has to be treated as an independent constant because we have no accurate theory for the proton magnetic moment in terms of other constants. The magnetic moment of the electron, on the other hand, is accurately predicted in terms of e, h, m_e and c by Dirac's theory of the electron plus the theory of quantum electrodynamics and does not have to be treated as a fundamental constant.

The least squares adjustment of fundamental constants

Many of the precision measurements made on physical systems can be related to the magnitudes of the SI units and the relevent fundamental constants when well tested theories are available. It is then possible to combine selected accurate results together with their individual standard errors in a least squares adjustment procedure that changes the values given in the constants until a set of 'best values' emerge, values which are consistent with the experimental results and the theories. This process was last carried out by Cohen (Cohen & Taylor 1973) and since then many research groups and specialist metrologists in standards laboratories have been developing a new generation of measurement techniques to provide input data for the evaluation of a new set of 'best values' by the international Committee on Data for Science and Technology (CODATA) task group on fundamental physical constants who expect to report their results at the end of 1983. Each analysis of this type yields a number of discrepancies which have to be resolved and in this process we improve our understanding of the physical world in the sense that the theories combined with the values of the constants provide a more consistent description of the experimental results.

We see, then, that the experimental determination of the fundamental constants involves the careful choice of accurate experiments which will influence favourably the least squares adjustment, for we are unable to make a direct precise measurement of a fundamental constant such as Planck's constant h or the proton mass m_p. The remainder of this paper will be concerned with some recent measurements of this type leaving others which involve the gravitational constant G, the constants associated with high energy physics and some aspects of cosmology to other papers in the conference.

Time, length and the velocity of light

More than 30 measurements of the velocity of light c, the constant that occurs so often in theories of relativity and electromagnetism, have been made since the earliest recorded experiment by Roemer in 1675. One of the more accurate experiments by Michelson in 1927 involved the direct measurement of the time taken for light pulses to travel a distance of 44 miles† and yielded an accuracy of the order $67/10^6$. Most of the measurements since 1950 have made use of the fact that c is equal to the product of frequency and wavelength for an electromagnetic wave in a

† 1 mile = 1.609 34 km.

vacuum. The most recent experiments by Evenson et al. in 1973 at the Bureau of Standards (U.S.A.) and Jolliffe et al. (1974) at the National Physical Laboratory (England) gave a much improved accuracy of $0.003/10^6$ because the measurements of both frequency and wavelength in vacuo were made simultaneously against the primary SI standards.

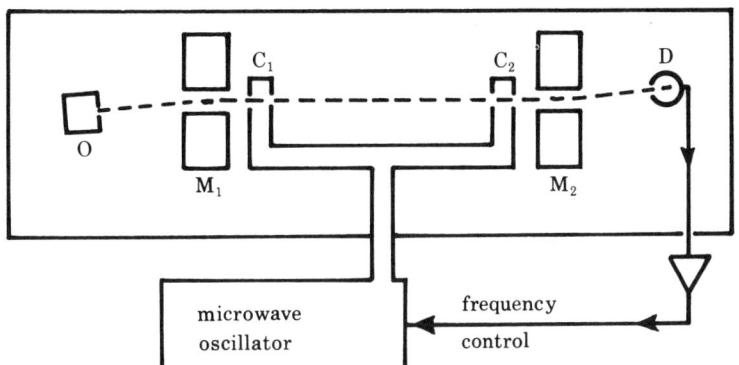

FIGURE 1. The caesium atomic clock. Caesium atoms leaving the oven O reach the detector D after passing through the inhomogeneous field focusing magnets M_1 and M_2 if they enter the microwave cavity C_1 in one hyperfine level W_1 and leave the cavity C_2 in the other level W_2. Resonance occurs when the microwave frequency f satisfies $hf = (W_1 - W_2)$ and the linewidth is $1/2T$ where T is the transit time from C_1 to C_2.

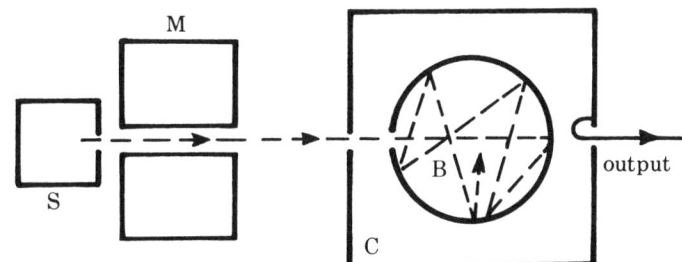

FIGURE 2. The hydrogen maser. Hydrogen atoms leaving the discharge source S in the upper hyperfine level W_1 are focused by the 4 or 6 pole magnet M into the coated bulb B in a microwave cavity C. Continuous oscillation at the hyperfine transition frequency $f = (W_1 - W_2)/h$ occurs when the beam intensity is increased until the induced downward transition rate to the lower level W_2 due to the microwave field already in the cavity is sufficient to make up the losses.

The SI primary standard of frequency and time is the atomic beam caesium clock, an apparatus in which a microwave oscillator is locked to the frequency $f = (W_1 - W_2)/h$ which maximizes the transition rate between two hyperfine levels W_1 and W_2 in the ground state of atomic ^{133}Cs moving without collisions in a vacuum. The second in SI units is formally defined as 9 192 631 770 periods of such an oscillator running under well defined conditions, a standard which has the amazing long term stability of 1 part in 10^{14} and an accuracy or reproduceability of 1 part in 10^{13}.

Well designed hydrogen masers oscillate at the microwave frequency associated with transitions between the ground state hyperfine levels of hydrogen atoms stored in a container. They have a better stability than the caesium clock over a period of a week to 10 days, about 1 part in 10^{15}, and they are more easily flown in satellites to carry out tests of relativity. They are not so good, however, as absolute time standards because the atoms collide with the walls of the container and the resultant reproduceability is only 1 part in 10^{12}.

The metre, the unit of length originally defined in terms of two scratch marks on the platinum–iridium bar kept at the Bureau International des Poids et Measures (B.I.P.M.) near Paris, was redefined in 1960 as 1 650 763.73 vacuum wavelengths of one of the orange–red lines in the spectrum of a ^{83}Kr lamp running under well defined conditions. The spacing between a pair of optically flat coated plates can be measured interferometrically in SI units with an accuracy of the order 1 part in 10^8, the limitation being mainly the linewidth and reproduceability of the krypton light source. The krypton standard for the metre was defined before the invention of the laser and the possibility of stabilized light sources in the infrared and visible with a short term stability approaching 1 part in 10^{12}, so we have every reason to expect a redefinition of the metre in the near future.

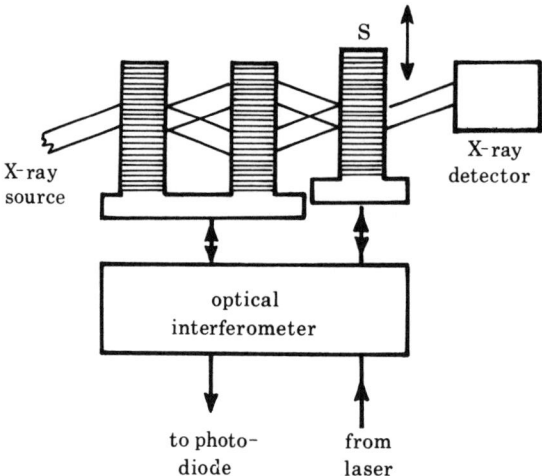

FIGURE 3. Measurement of a lattice plane separation in silicon. Movement of the silicon crystal S in the direction shown produces a pattern with period d equal to the lattice spacing from the X-ray detector and a pattern with a period equal to half an optical wavelength from the optical detector. Careful analysis of the patterns recorded simultaneously as the crystal moves allows a direct estimate of d in terms of the wavelength of the light.

The technique used at N.P.L. to measure the speed of light involved a direct comparison of the wavelengths of the standard line from the krypton lamp and radiation at a similar frequency of 5×10^{14} Hz obtained by mixing light from several highly stabilized lasers. The laser frequencies were measured by complicated beating techniques in terms of the atomic frequency standard at about 10^{10} Hz and the resultant speed of light deduced from the frequency wavelength product had an error limited mainly by the linewidth of the krypton lamp.

Because of the uncertainty in any measurement of a length in terms of the krypton standard it is possible that the next meeting of the Conference General des Poids et Measures in 1983 may define the metre in terms of the velocity of light as the distance travelled in a vacuum in the time 1/2 997 924.58 s measured in terms of the caesium clock. This definition will keep the magnitudes of the length and time standards at their present 'best values' and allow frequency stabilized lasers to be used as local transfer standards to obtain more precise length measurements.

Mass

The present mass standard is related to the international prototype of the kilogram kept at B.I.P.M. in France and local standards calibrated by direct weighing to about $0.01/10^6$ are kept in the national standards laboratories. Although atomic masses can be compared in mass spectrometers with a precision approaching $0.01/10^6$ an atomic mass standard cannot be used at present because the Avogadro number, which gives the number of atoms in a mass of material equal to the atomic number in grams, is not known with sufficient precision. This situation may change in the near future since Becker *et al.* (1981) showed that it is possible to make an accurate measurement of the lattice spacing in a pure perfect crystal of ^{26}Si by using X-ray diffraction and optical interference to measure the number of lattice spacings in a half wavelength of light. This will allow the mass standard to be redefined in terms of the mass of a silicon crystal of known dimensions and leave us with standards of mass, length and time in terms of natural atomic systems.

Electrical standards

Electronic techniques play a crucial role in any precision measurement so it is important that the accuracy of the electrical standards continue to be improved beyond the $1/10^6$ level typically available in well equipped laboratories. A very significant advance has been made in this direction in the last 20 years by the realization that the Josephson effect (Josephson 1962) can be used to define the volt in terms of the unit of time. It is found that a superconducting current passes by tunnelling across a gap as small as 1 nm separating two superconductors when the voltage V across the gap and the frequency f of microwave radiation applied are given by $2eV = nhf$, where n is an integer and h is Planck's constant. The ratio $2e/h$ measured in this way is found to be independent of the materials used to form the junction to better than $0.001/10^6$ (Gallop 1928) and most standards laboratories are now using the technique to maintain the volt in terms of $2e/h$ with a precision of about $0.01/10^6$. Details of the way this is done are described in recent reviews by Petley (1980) and Gallop (1982).

The accurate measurement of magnetic field is usually carried out by a magnetic resonance technique involving the protons in water or some other system with magnetic moment μ and spin J. Transitions between the states in the field B are detected when the frequency f of an additional oscillating field is given by $hf = \mu B/J$, so it is important that the relevent nuclear and atomic magnetic moments are available with high precision. Measurements such as the ratio of the proton resonance frequency to the proton cyclotron circulation frequency in a given field as well as the magnitude of the proton moment in hertz per tesla provide important data for the least squares adjustment of the fundamental constants and also lead to an accurate value for the fine structure constant α.

The fine structure constant

The fine structure constant $\alpha = e^2 \mu_0 c/2h$ is a particularly important dimensionless combination of fundamental constants with the value $1/137.035\,965$ and uncertainty $0.09/10^6$. The fine structure splitting in atoms and many quantum electrodynamic expressions for quantities such as the electron g_e factor and the proton gyromagnetic ratio can be expressed as a power series in α. The electron and positron g factors have recently been measured (Schwinberg *et al.* 1981) with an accuracy of 4 parts in 10^{11} by studying the behaviour of single electrons and positrons stored in an electromagnetic trap. They are probably the most accurate measurements of any property of a fundamental particle made so far but the value of α^{-1} deduced from the results is $0.33 \pm 0.14/10^6$ greater than that derived from an accurate measurement of the proton gyromagnetic ratio (Williams & Olsen 1979), a discrepancy which may imply a need for a higher power of α in the expression for g_e. The fine structure constant has also been deduced by Tsui *et al.* (1982) with an accuracy of $0.09/10^6$ by measuring the quantized Hall resistance of a gallium arsenide hetrostructure in which a two dimensional electron gas is formed and the result is in good agreement with the proton gyromagnetic ratio value.

Conclusions

The last 20 years have seen a steady change over from standards like the standard cell and the quartz clock, which required very careful control during manufacture and an extremely stable environment to obtain good precision, to standards based on well understood quantum systems. Already the second depends on the atomic caesium clock and the volt is maintained by observing Josephson junction voltage steps. Before long it is likely that the metre will be defined in terms of an adopted value for the speed of light and the caesium second, the kilogram will be related to an improved Avogadro number based on the lattice spacing in a crystal, and the ohm may well be defined in terms of quantized Hall resistance. It will then be much easier to maintain precise standards in different research laboratories around the world and this in turn will have a beneficial effect on the tests of basic theories and the determination of the fundamental constants, developments that must necessarily proceed together if the least squares adjustment technique is used to determine 'best values' for the constants. We cannot predict at the moment whether the so called 'fundamental constants' will change with time as the Universe expands but there is little doubt that interesting new discoveries and theories will emerge as the precision improves.

References

Becker, P., Dorenwendt, K., Ebeling, G., Lauer, R., Lucas, W., Probst, R., Rademacher, H. J., Reim, G., Seyfried, P. & Siegert, H. 1981 *Phys. Rev. Lett.* **46**, 1540–1543.
Cohen, E. R. & Taylor, B. N. 1973 *J. phys Chem. Ref. Data* (no. 4), 663–690.
Gallop, J. C. 1982 *Metrologia* **18**, 67–92.
Jolliffe, B. W., Rowley, W. R. C., Shotton, J. C., Wallard, A. J. & Woods, P. T. 1974 *Nature, Lond.* **251**, 46–47.
Josephson, B. D. 1962 *Physics Lett.* **1**, 251–255.
Petley, B. W. 1980 *Metrology and the fundamental constants* (ed. A. Ferro Milone & P. Giacomo), pp. 358–463. Amsterdam: North-Holland.
Schwinberg, P. B., Van Dyck, R. S. Jr & Dehmelt, H. G. 1981 *Phys. Rev. Lett.* **47**, 1679–1682.
Tsui, D. C., Gossard, A. C., Field, B. F., Cage, M. E. & Dziuba, R. F. 1982 *Phys. Rev. Lett.* **48**, 3–6.
Williams, E. R. & Olsen, P. T. 1979 *Phys. Rev. Lett.* **42**, 1575–1579.

Towards the next evaluation of the fundamental physical constants†

By B. W. Petley

*Division of Quantum Metrology, National Physical Laboratory,
Teddington, Middlesex TW11 0LW, U.K.*

The note discusses the experimental data likely to be considered in the forthcoming evaluation of the 'best' values of the fundamental physical constants by Cohen and Taylor, and the implications for the realization of the ampere.

As far as is known, the accurately measured fundamental physical constants show no significant variation with time when expressed in terms of dimensionless quantities. The situation is rather more complicated when the experimental values are expressed in SI units, and increasingly reflects our inability to realize the definitions of the SI units with adequate precision. From some points of view this is a happy state of affairs for it allows us to reframe our definitions of the SI units in terms of the fundamental constants: as with the new definition of the metre (Petley 1983), or the use of the Josephson effect value of $2e/h$, E_J, to maintain the volt.

The CODATA task group on fundamental constants, of which the author is a member, has been considering the experimental data that will probably be included in the forthcoming evaluation by Cohen and Taylor. In this consideration, it is quickly apparent that the accuracy of K (the ratio of the maintained ampere to the 'absolute' ampere of the SI definition), has become of increasing importance since the 1969 review by Taylor, Parker and Langenberg. Thus the estimates of the values of such constants as the elementary charge, e, the electron mass m_e, the proton mass m_p, and the Planck constant h, all depend critically on the value of $K(e)$ or K^2 (the remainder). A further limitation is provided by the measurements of the fine structure constant α.

Cohen (1981) has shown that the crucial experiments can be displayed graphically as a Birge–Bond diagram by expressing them in terms of more accurately determined constants (such as the Rydberg constant, μ'_p/μ_B, E_J, c, and the ratio of the maintained ohm to absolute SI ohm, $\bar{R}$), and two unknowns: K and the fine structure constant α. This simplification is partly allowed by the improved precision of the direct measurements of m_p/m_e by the ion-trap technique developed at the University of Washington (Van Dycke & Schwinberg 1981). Their result is in any case in good agreement with other direct measurements (see review by Wineland *et al.* 1983), and with the indirect measurements of μ'_p/μ_N of earlier times (Mamyrin *et al.* 1972; Petley & Morris 1974), combined with the Phillips *et al.* (1975) value of μ'_p/μ_B. The Rydberg constant provides an important auxiliary constant and the most accurate of the new generation of laser spectroscopy measurements is that of Amin *et al.* (1981). This has an accuracy comparable with the reproducibility of the krypton-86 realization of the metre and is in reasonable agreement with the measurements of lesser precision by Goldsmith *et al.* (1978) and Petley *et al.* (1980).

The present results are displayed in figure 1, with K and $\alpha^{-1}K$ as horizontal and vertical

† This paper is an amplified version of a contribution to the discussion, invited by the chairman, following Professor Smith's paper.

axes, and a range of values for the coordinates of only a few parts per million. Ideally all of the experimental measurements should be consistent with a unique value of α^{-1} and K, within the assigned uncertainties. On this plot measurements of the silver Faraday F (N.B.S.: Bower & Davis 1980), and the gyromagnetic ratio of the proton in a strong magnetic field, γ'_p(high) (N.P.L.: Kibble & Hunt 1979; N.I.M.–P.R.C.: Chiao et al. 1980; Wang 1981) determine $\alpha^{-1}K$, and generate a horizontal line.

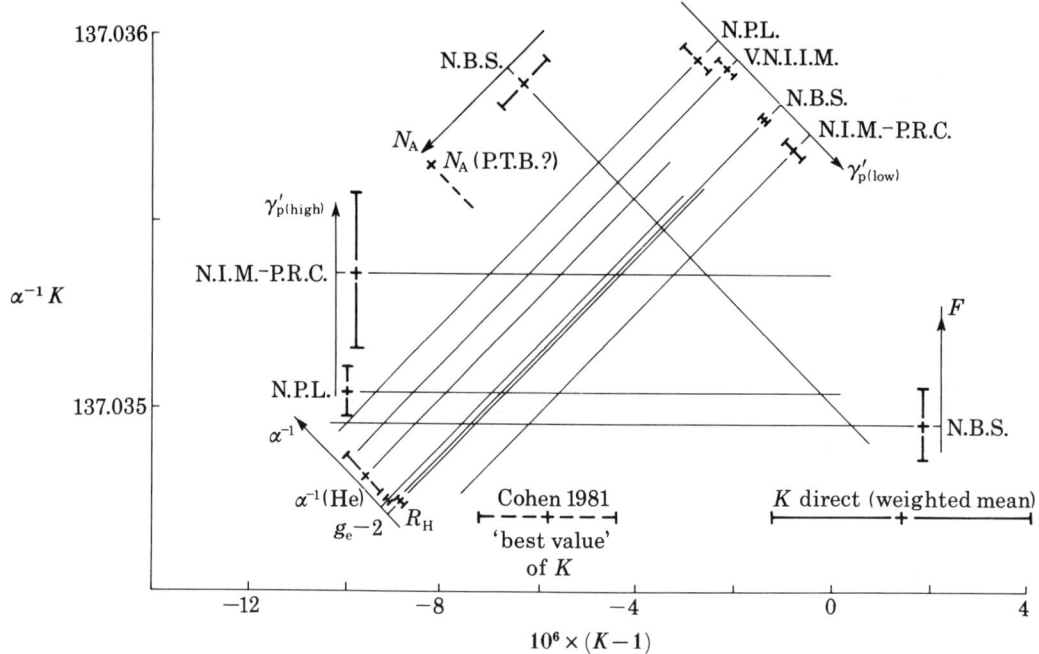

FIGURE 1. Birge–Bond display of most of the present data relating to the fine structure constant α, and ampere conversion constant K (i.e. ampere maintained at the B.I.P.M./ampere SI); each result has an associated standard deviation uncertainty and generates an appropriate line, as shown (based on the figure and data in Cohen 1981). The national metrological laboratories identified here by initials are given in full in the Appendix.

The direct realizations of the ampere K, yield a vertical line and these have been represented by the weighted mean of the available results (see Cohen & Taylor 1973; Cohen 1981). Elnekave & Fau (1981) have reported a preliminary absolute volt realization. Although their work, as with that of others (see, for example, Kibble et al. 1983), is not yet finally completed, the results are tending to confirm the estimates of K obtained via the fundamental constants.

The Avogadro constant, N_A(N.B.S.), has been measured by Deslattes et al. (see Deslattes 1980) by measuring the density and lattice parameter of very pure samples of silicon. Their measurement yields $\alpha^{-1}K^2$, defining a rectangular hyperbola on the figure. Incidentally the P.T.B. measurement of the lattice parameter of silicon, if confirmed by a corresponding value for N_A (N_A(P.T.B?): Becker et al. 1981) would give a value in better accord with the γ'_p(high) and Faraday measurements.

There are two remaining groups of results which essentially determine α^{-1}. The non-QED value is obtained via the measurements of the gyromagnetic ratio of the proton in a weak magnetic field, γ'_p(low). The most precise of these is that of Williams & Olsen (1979) (N.B.S.). Other measurements have been reported by Tarbeyev (1981) (V.N.I.I.M.), by Vigoureux & Dupuy (1980) (N.P.L.), and by Wang (1981) (N.I.M.–P.R.C.). The other determinations of

α^{-1} tend to involve QED, with the possible exception of the measurements of the quantized Hall resistance, R_H. The latter was first demonstrated as a measurement of α^{-1} by Klitzing et al. (1980). More accurate measurements by this technique have already been reported by Yamanouchi et al. (1981), by Tsui et al. (1982), and by Bliek et al. (1983). At present there are some discrepancies at the part per million level (possibly in the measurement process) and only one result has been shown for clarity. The most accurate estimate of the fine structure constant by a spectroscopic technique now comes from the measurements of Kponou et al. in helium (α^{-1}(He): Hughes 1981), as a result of improvements in the theory. The remaining, and most accurate estimate of α, is obtained from the determination of the g-factor anomaly of the free electron ($g_e - 2$) at the University of Washington by Van Dycke et al. (1979).

The data pertaining to the fine structure constant is reasonably consistent with the 1973 recommended value and leads to the conclusion that all continues to be well with quantum electrodynamic calculations to the nth decimal place. The situation regarding the best value of K shows little change since that discussed by Taylor (1976). The results are not as consistent as one would like, but they suggest that the 'best value' would be about 6 parts per million less than unity, as shown (the 1973 data was compatible with unity). There are likely to be consequent changes (ca. $2\sigma_{73}$) in the recommended values of many of the fundamental constants in the forthcoming evaluation (meanwhile the 1973 values should continue to be used). Of the remaining constants, the Boltzmann constant may be obtained from measurements of the gas constant (Quinn et al. 1976) and there is a new measurement of the gravitational constant, Luther et al. (1982), of precision better than 0.01 % that will supersede the 1973 recommended value. The forthcoming evaluation will be rather more sophisticated than the simplified analysis given here; thus the ratio of the maintained ohm to the absolute SI ohm is likely to be evaluated as well. Evidently, there is a strong incentive for the experimenter and theoretician to strive for improvement in the measurements, despite the fact that there is an important group of users who, quite sensibly, approximate the velocity of light as 3×10^8 m s^{-1}.

References

Amin, S. R., Caldwall, C. D. & Litchen, W. 1981 *Phys. Rev. Lett.* **47**, 1234.

Becker, P., Dorenwendt, K., Ebeling, G., Lauer, R., Lucas, W., Probst, R., Rademacher, H.-J., Reim, G., Seyfried, P. & Seigert, H. 1981 *Phys. Rev. Lett.* **46**, 1540.

Bliek, L., Braun, E., Englemann, H.-J., Leontiew, H., Melchert, F., Schlapp, W., Stahl, B., Warnecke, P. & Weimann, G. 1983 *P.T.B.-Mitt.* **93**, 21.

Bower, V. E. & Davies, R. S. 1980 *J. Res. natn Bur. Stand.* **85** (3), 175.

Chiao, W., Lui, R. & Shen, P. 1980 *IEEE Trans. Instrum. Meas.* **IM-29**, 238.

Cohen, E. R. 1981 In *Quantum metrology and fundamental constants* (ed. P. H. Cutler & A. A. Lucas) NATO-ASI ser. New York: Plenum. (In the press.)

Cohen, E. R. & Taylor, B. N. 1973 *J. phys. Chem. ref. Data* **2**, 663.

Deslattes, R. D. 1980 In *Metrology and fundamental constants*. Proceedings 68th Int. School Enrico Fermi (ed. A. Ferro Milone, P. Giacomo & S. Leschuitta), pp. 38–114. New York: North Holland.

Elnekave, N. & Fau, A. 1981 In Proceedings of the Second International Conference on Precision Measurements and Fundamental Constants (ed. B. N. Taylor & W. D. Phillips). *Natn Bur. Stand. spec. publ.* 617. Washington: U.S. Government printing office. (In the press.)

Goldsmith, J. E. M., Weber, E. W. & Hansch, T. W. 1978 *Phys. Rev. Lett.* **41**, 1525.

Hara, K., Shiota, F. & Kubota, T. 1981 In Proceedings of the Second International Conference on Precision Measurements and Fundamental Constants (ed. B. N. Taylor & W. D. Phillips). *Natn Bur. Stand. spec. publ.* 617. Washington: U.S. Government printing office. (In the press.)

Hughes, V. W. 1981 In Proceedings of the Second International Conference on Precision Measurements and Fundamental Constants (ed. B. N. Taylor & W. D. Phillips). *Natn Bur. Stand. spec. publ.* 617. Washington: U.S. Government printing office. (In the press.)

Kibble, B. P. & Hunt, G. J. 1979 *Metrologia* **15**, 5.

Kibble, B. P., Smith, R. C. & Robinson, I. A. 1983 *IEEE Trans. Instrum. Meas.* **IM-13**, 141.
Klitzing, K. V., Dorda, G. & Pepper, M. 1980 *Phys. Rev. Lett.* **45**, 494.
Luther, G. G. & Towler, W. R. 1982 *Phys. Rev. Lett.* **48**, 121.
Mamyrin, B. A., Aruyev, N. N. & Alekseenko, S. A. 1972 *Zh. èksp. teor. Fiz.* **63**, 3. (Eng. transl.: *Soviet Phys. JETP* **36**, 1 (1973).)
Petley, B. W. 1983 *Nature, Lond.* **303**, 373.
Petley, B. W. & Morris, K. 1974 *J. Phys.* B **7**, 167.
Petley, B. W., Morris, K. & Shawyer, R. E. 1980 *J. Phys.* B **13**, 3099.
Phillips, W. D., Cooke, W. E. & Kleppner, D. 1975 *Phys. Rev. Lett.* **35**, 1619.
Quinn, T. J., Colclough, A. R. & Chandler, T. R. D. 1976 *Phil. Trans. R. Soc. Lond.* A **283**, 367.
Quinn, T. D. & Martin, J. S. 1981 In Proceedings of the Second International Conference on Precision Measurements and Fundamental Constants (ed. B. N. Taylor & W. D. Phillips). *Natn Bur. Stand. spec. publ.* 617. Washington: U.S. Government printing office. (In the press.)
Tarbeyev, Yu. V. 1981 In Proceedings of the Second International Conference on Precision Measurements and Fundamental Constants (ed. B. N. Taylor & W. D. Phillips). *Natn Bur. Stand. spec. publ.* 617. Washington: U.S. Government printing office. (In the press.)
Taylor, B. N. 1976 *Metrologia* **12**, 86.
Tsui, D. C., Gossard, A. C., Field, B. F., Cage, M. E. & Dziuba, R. F. 1982 *Phys. Rev. Lett.* **48**, 3.
Van Dycke, R. S. & Schwinberg, P. B. 1981 *Phys. Rev. Lett.* **47**, 395.
Van Dycke, R. S., Schwinberg, P. B. & Dehmelt, H. G. 1979 *Bull. Am. phys. Soc.* **24**, 758.
Vigoureux, P. & Dupuy, N. 1980 N.P.L. report DES59.
Wang, Zhu-Xi 1981 In Proceedings of the Second International Conference on Precision Measurements and Fundamental Constants (ed. B. N. Taylor & W. D. Phillips). *Natn Bur. Stand. spec. publ.* 617. Washington: U.S. Government printing office. (In the press.)
Williams, E. R. & Olsen, T. 1979 *Phys. Rev. Lett.* **42**, 1575.
Wineland, D. J., Itano, W. M. & Van Dycke, R. S. 1983 In *Advances in atomic and molecular physics* **19** (ed. B. Bederson & D. R. Bates). New York: Academic Press. (In the press.)
Yamanouchi, C., Yoshihiro, K., Kinoshita, J., Inagaki, K., Moriyama, J., Baba, S., Kawaji, S., Murakami, K., Igarashi, T., Endo, T., Koyanagi, M. & Nakamura, A. 1981 In Proceedings of the Second International Conference on Precision Measurements and Fundamental Constants (ed. B. N. Taylor & W. D. Phillips). *Natn Bur. Stand. spec. publ.* 617. Washington: U.S. Government printing office. (In the press.)

APPENDIX. EXPLANATION OF ABBREVIATIONS IN FIGURE 1.

N.B.S.: National Bureau of Standards, U.S.A.;

N.I.M.-P.R.C.: National Institute of Metrology, People's Republic of China;

N.P.L.: National Physical Laboratory, U.K.;

P.T.B.: Phisikalische-Technische Bundesanstalt, F.R.G.;

V.N.I.I.M.: All-Union Scientific Research Institute of Metrology (Mendeleev Institute), U.S.S.R.

The search for proton decay and other rare phenomena

By M. Goldhaber

Brookhaven National Laboratory, Upton, New York 11973, U.S.A.

We have obtained the first results with a large water Cherenkov counter located in a salt mine in Ohio. (Bionta *et al.* 1983. Submitted to *Phys. Rev. Lett.*) The partial lifetime for the decay $p \to e^+\pi^0$ is found to be longer than the value calculated from the minimal SU(5) theory, with the best presently available estimates for the parameters needed. The detector is sensitive to some other potential decay modes, as well as to some rare phenomena that have been discussed in recent years. The characteristics of the penetrating cosmic ray muons and atmospheric neutrino interactions observed are compatible with expectations.

Astronomers are fond of saying, 'the absence of evidence is not the evidence of absence'. Nevertheless, it was believed for a long time that protons are absolutely stable. Since 1954, however, proton stability has been subjected directly to fairly sensitive empirical tests (Reines *et al.* 1954).

Sometimes I am asked how my interest in proton decay started. As I remember it, it began about 30 years ago when the theory of continuous creation advocated by Bondi, Gold and Hoyle was much discussed. If protons could be created out of nothing, might they not also have a chance to disappear? If such a disappearance took place, it might show up by leaving a nuclear excitation. This could be detected, for instance, in thorium by what would look like spontaneous fission. But once one considers disappearance of protons, a more conservative approach would be to ask whether protons could decay into other particles conserving energy, momentum, electric charge, and angular momentum. At that time, Reines and Cowan had a large neutrino detector, a scintillation counter, and it seemed natural to use it to search for charged particles which might arise from proton decay. Thus, the first proton decay experiments and many thereafter were parasitic to a neutrino search. Nowadays, this has often been reversed, with dedicated proton decay detectors also being used as neutrino detectors; these experiments are now symbiotic, rather than parasitic.

Over the years, the limits on proton lifetime were increased by various methods to around 10^{30} years. Then, starting in 1973 with suggestions by Pati and Salam, by Georgi and Glashow, and by Georgi, Quinn and Weinberg, and by others, the idea of grand unification was shown to lead to the prediction that quarks could change into leptons, making protons unstable. In particular, the so-called minimal SU(5) theory, and equivalently some higher groups, for example, SO(10) and SU(16), as emphasized especially by Weinberg and Pati, make a fairly well defined prediction for the proton lifetime, not much longer than the previous experimental limit. It seemed therefore possible to test this prediction with reasonably sized detectors.

Many such experiments are now in progress or being prepared. The Kolar gold field experiment found three contained events, considered as candidates for proton decay, and the Mont Blanc experiment has reported one candidate. The Irvine–Michigan–Brookhaven experiment (Bionta *et al.* 1983) has recently reported its first results:

'Observations were made 1570 mwe underground with an 8000 metric ton water Cherenkov detector. During a live-time of 80 days no events consistent with the decay p → e$^+\pi^0$ were found in a fiducial mass of 3300 metric tons. We conclude that the limit on the lifetime for bound plus free protons divided by the e$^+\pi^0$ branching ratio is $\tau/B > 6.5 \times 10^{31}$ years; for free protons our limit is $\tau/B > 1.9 \times 10^{31}$ years (90% confidence). Observed cosmic ray muons and neutrinos are compatible with expectations.'

While we concentrated first on the e$^+\pi^0$ decay mode, we are also sensitive to most other potential two-body decay modes, as well as to some multi-body modes. By Monte Carlo simulation, the sensitivity of our detector for each decay mode, as well as the neutrino induced background of similar appearance, can be estimated. In physics, reproducibility is vital. With rare events we must be very careful: Nature might be giving us a Rorschach test! We can give limits for the K$^0\mu^+$ decay mode ($> 1.4 \times 10^{31}$ a) as well as for n–n̄ oscillations in oxygen ($> 10^{31}$ a). According to Dover et al. (1983), this corresponds to a free n–n̄ oscillation time of greater than 5×10^7 s.

We have also carried out a search (Errede et al. 1983) for magnetic monopoles 'catalyzing' nucleon decay as suggested by Rubakov and Callan.

'No positive evidence for such a process has been found during 100 days of detector livetime. For an average cross section of H$_2$O > 10 mb† and velocities $10^{-4} < \beta_m < 10^{-1}$, we find a limit for the monopole flux $< 7.2 \times 10^{-15}$ cm^{-2} sr^{-1} s^{-1}.'

References

Bionta, R. M., Blewitt, G., Bratton, C. B., Cortex, B. G., Errede, S., Foster, G. W., Gajewski, W., Goldhaber, M., Greenberg, J., Haines, T. J., Jones, T. W., Kielczewska, D., Kropp, W. R., Learned, J. G., Lehmann, E., LoSecco, J. M., Ramana Murthy, P. V., Park, H. S., Reines, F., Schultz, J., Shumard, E., Sinclair, D. Smith, D. W., Sobel, H. W., Stone, J. L., Sulak, L. R., Svoboda, R., van der Velde, J. C. & Wuest, C. 1983 (Submitted to *Phys. Rev. Lett.*)

Dover, C. B., Gal, A. & Richard, J. M. 1983 *Phys. Rev.* D **27**, 1090–1100.

Errede, S., Stone, J. L., van der Velde, J. C., Bionta, R. M., Blewitt, G., Bratton, C. B., Cortez, B. G., Foster, G. W., Gajewski, W., Goldhaber, M., Greenberg, J., Haines, T. J., Jones, T. W., Kielczewska, D., Kropp, W. R., Learned, J. G., Lehmann, E., LoSecco, J. M., Murthy, P. V. R., Park, H. S., Reines, F., Schultz, J., Shumard, E., Sinclair, D., Smith, D. W., Sobel, H. W., Sulak, L. R., Svoboda, R. & Wuest, C. 1983 (Submitted to *Phys. Rev. Lett.*)

Reines, F., Cowan, C. L. Jr & Goldhaber, M. 1954 *Phys. Rev.* **96**, 1157–1158.

† 1b = 10^{-28} m^2.

The constancy of G and other gravitational experiments

By R. D. Reasenberg

Radio and Geoastronomy Division, Smithsonian Astrophysical Observatory, Harvard-Smithsonian Center for Astrophysics, Cambridge, Massachusetts 02138, U.S.A.

Traditionally, theories of gravitation have received their most demanding tests in the solar-system laboratory. Today, electronic observing technology makes possible solar-system tests of substantially increased accuracy. We consider how these technologies are being used to study gravitation with an emphasis on two questions:

(i) Dirac and others have investigated theories in which the constant of gravitation, G, appears to change with time. Recent analyses using the Viking data yield $|\dot{G}/G| < 3 \times 10^{-11}$ per year. With further analysis, the currently available ensemble of data should permit an estimate of $\dot{G}/G$ with an uncertainty of 10^{-11} per year. At this level it will become possible to distinguish among competitive theories.

(ii) Shapiro's time-delay effect has provided the most stringent solar-system test of general relativity. The effect has been measured to be consistent with the predictions of general relativity with a fractional uncertainty of 0.1%. An improved analysis of an enhanced data set should soon permit an even more stringent test.

Technology now permits new kinds of tests to be performed. Among these are some that measure relativistic effects due to the square of the (solar) potential and others that detect the Earth's 'gravitomagnetic' field (the Lense–Thirring effect). These experiments, and the use of astrophysical systems are among the experimental challenges for the coming decades.

1. Introduction

The energetic interplay of theory and experiment has led to the rapid advance of our understanding of the physical universe. In no other field has our knowledge of reality evolved so rapidly. For many decades it appeared that general relativity was to be an exception to this rule. Now, thanks to the application of modern technology, gravitational physics is again an experimental subject. Most tests today are limited to the first-order post-Newtonian effects and have, at best, an accuracy of a part in 10^3. Of all the reputable experiments done, there has not been one that yielded results inconsistent with the predictions of general relativity. Over the next decades, the evolving level of technology applied to these tests will lead to order-of-magnitude increases in the accuracy of the results. But we must ask ourselves whether we should not be looking at alternative régimes for testing this theory of gravitation. The identification of new physically realizable experiments is a significant frontier for relativity today.

The classical laboratory for gravity research is the solar system, where the theories of Newton and Einstein have been tested in turn. Although certain observations are particularly important to a particular 'test', the entire ensemble of data is advantageously used for each of the possible tests. In the traditional laboratory environment, an apparatus is assembled and operated in such a manner as to obtain an estimate of a single or, at most, a few parameters of interest. These are generally described as physical quantities. The contrivances of a terrestrial laboratory are precluded in a solar-system study. Here an ensemble of diverse data is collected and a multi-parameter model is fitted to the data. The vast majority of the parameters estimated are of no interest to the relativist. These 'nuisance' parameters, however, are essential to the reliable determination of those few quantities that are of relativistic interest.

The study of post-Newtonian effects in the solar system is difficult. They are scaled by the solar potential which reaches 2×10^{-6} at the limb of the Sun, a region accessible to the experimenter only by means of photons. As will be seen below, the solar-system laboratory is useful only because of the availability of modern technologies of electronics and space travel. The natural alternative, terrestrial laboratory experiments, is not yet practical because of the small strength of the gravitational interaction between laboratory bodies and the high level of ambient noise, both natural and man made. Finally, there is the possibility of using astrophysical systems as serendipitous laboratories. For such systems, the gravitational potential, ϕ, may be large. However, these systems are remote, unfamiliar, and generally observed by a single means. Thus, degeneracy and model ambiguity are major problems. The pulsar in a binary system, PSR $1913+16$, is a prime example of an astrophysical system that yields information on the nature of the laws of gravitation.

TABLE 1. SOME RELATIVISTIC EFFECTS

effect	parametric dependence	state of determination present	expected: σ	proposed: σ
deflexion of light	$\frac{1}{2}(1+\gamma)$	1.007 ± 0.009	0.001	10^{-6}
perihelion advance	$\frac{1}{3}(2+2\gamma-\beta)$	0.98 ± 0.04	0.003	—
gravitational redshift	$\dfrac{\text{observed}}{\text{expected}}$	1.0000025 ± 0.00014	—	10^{-8}
Shapiro time delay	$\frac{1}{2}(1+\gamma)$	1.000 ± 0.001	0.0005	—
Nordtvedt's principle of equivalence violation	$\eta = 4\beta-\gamma-3$	0.0 ± 0.015	0.005	—
secular variation of G	$\dot{G}/G$	$(0 \pm 3) \times 10^{-11}/\text{a}$	$10^{-11}/\text{a}$	—
gravitational radiation damping of orbit	$\dfrac{\text{observed}}{\text{expected}}$	0.957 ± 0.09	—	—
geodetic precession	$\frac{1}{3}(1+2\gamma)$	—	—	0.001
Lense–Thirring effect*	$\frac{1}{2}(1+\gamma)$	—	—	0.1

* The significance of a test of this effect transcends the importance of determining the p.p.n. coefficient. The detection by the Stanford gyroscope experiment of the Lense–Thirring effect would be the first experimental verification of the relativistic effect of the (rotational) motion of the source.

Tests of general relativity, like those of any theory, are often described in terms of 'effects'. These may be thought of as predictions or observations that differ from those expected according to a previous description of the laws of nature. Table 1 lists some effects that have been or will be important in tests. As a list of known relativistic effects, it is far from complete. For example, it excludes variations of the gravitational constant with distance on a laboratory scale, gravitational waves, and quantum gravity effects.

In the following sections, we will consider first the ensemble of data under analysis at the Center for Astrophysics (C.f.A.). This work has been in collaboration with R. W. Babcock, J. F. Chandler and I. I. Shapiro at C.f.A. and with R. W. King at M.I.T. We will then discuss areas of relativistic interest, the secular variation of G and the Shapiro time-delay test. Finally, §5 will be devoted to a discussion of experiments currently under consideration. Some of these are well developed and ready for implementation; others are in an early phase of a long process and have an uncertain future.

2. SOLAR SYSTEM DATA SET

Before considering specific tests, we review the current state of solar-system analysis at C.f.A. Table 2 shows our present data set. For current purposes, the most powerful data are the Viking

Lander observations, of which we use about 1060, spanning 6 years. In addition, we have 1075 Doppler observations of the Viking Landers and 4060 Viking Orbiter normal points (n.p.). Although these last two kinds of data have little direct value for the relativity tests, the former are useful for determining the rotation of Mars, and the latter establish the mean distance of the Landers from the Mars equator. Since these are part of the solar-system modelling problem, the Doppler and n.p. data are indirectly useful for the relativity analysis. The Mariner-9 mission yielded spacecraft orbiter normal points for 1971 and 1972. Although there are fewer than 200 of these, they are important in determining the Mars ephemeris because they have epochs long before the Viking era. The table shows no Mars radar data, though we do have 2.5×10^4 of these, all of which we have previously processed. The combination of Viking and Mariner-9 data is so much more powerful that we temporarily discarded the Mars radar data.

TABLE 2. COMBINED SETS OF DATA

source†	no. of data	approximate range of error assumed in estimator		unit
		min	max	
Viking				
Lander delay (plasma corrected)	798	20	60	ns
Lander delay (not plasma corrected)	263	50	300	ns
Orbiter n.p.‡	4060	100	900	ns
Lander Doppler	1075	20	40	mHz
l.l.r.				
Observing session n.p.§	2613	6	14	ns
Mariner 9				
Orbiter n.p.‡	185	0.1	10	μs
radar				
Mercury	642	1	15	μs
Venus	784	1	15	μs
meridian circle‖				
Sun	1023	≈ 2		″
Moon	212	≈ 0.5		″
inner planets (M, V, M)	1518	≈ 1		″
outer planets (J, S, U, N)	1643	≈ 1		″
outer planet n.p.¶	6	25	500	μs

† All observables are time delays except for the Viking Lander Doppler and meridian circle data.

‡ The orbiter normal point (n.p.) is a compressed datum: the equivalent Earth–Mars time delay measured between the centres of mass of the planets.

§ The lunar laser ranging (l.l.r.) normal point (n.p.) is a single estimate of the round trip propagation time between a tracking station and a single lunar retroreflector. The estimate is an average based on all photons received during an observing sequence.

‖ The data are a mixture of right ascension and declination measurements.

¶ The outer planet normal point (n.p.) is a compressed datum from a spacecraft encounter with either Jupiter or Saturn. The n.p. is the equivalent Earth–planet time delay measured between the centres of mass of the planets.

Lunar laser ranging has given us 2600 so-called laser ranging normal points. Traditionally, the lunar data were analysed separately from the planetary data, in part because the scientific basis for the work is different and in part because different subgroups within the research group had been doing the analyses.

In principle, the combination of the lunar and planetary data sets should have been a simple matter as each had been previously reduced to normal equations. Even though disparate nominal values were used for the adjustable parameters in the two reductions, a simple linear operation, which our software will perform, should properly combine the data. In practice, the problem is more complicated. The estimator shows large (0.9999) correlations that tend to enhance the undesirable effects of both estimator nonlinearity and (economically necessary) approximations that cause errors in the variational equations. The common (and economically justified) practice of re-using previously calculated variational equations exacerbates this problem. It therefore is necessary to iterate the estimator in order to combine properly the data. At the same time, the combined data set may require an enhanced model. Thus there are two competing lines of activity that must be pursued in balance to achieve a properly combined data set.

It now appears that we have solved the most serious of the problems and successfully have combined the data sets. To remove remnants of the separate histories of the two data groups required four iterations of the estimator, the last of which was distinguished in that we recomputed the variational equations.

3. Secular variation of G

Dirac (1937, 1938) has investigated the cosmological consequences of the large numbers hypothesis. It was noted even earlier that it is possible to combine physical constants to create dimensionless numbers that generally differ from unity by at most a few orders of magnitude. Yet there are some such numbers that are extremely large. One of these is the ratio of the electric to the gravitational force between an electron and a proton; it is close to 10^{40}. Another is the age of the universe expressed in units of atomic time; it is also close to 10^{40}. Finally there is the mass of the visible universe expressed in proton masses; it is close to $10^{(40 \times 2)}$. The hypothesis is that this coincidence is a message, not an accident; perhaps these quantities are related by some small time-invariant constants. If that is correct, then one or more of the 'constants' used to make each of the large numbers must be time varying. The original description was that the gravitational 'constant' was the most likely candidate for the time variable. A discussion of some alternatives is given by Dyson (1972).

How would one detect such a variable G in the solar system? We should have a complete theory of dynamics with a variable G, yet none exists. As an experiment, based on Dirac's early papers, we use an *ad hoc* model (Shapiro 1971)

$$G = G_0 + \dot{G}(t - t_0), \tag{1}$$

where t_0 is a convenient recent epoch. This is not an entirely satisfactory way to describe the dynamics, but it allows us to determine whether or not there may be a secular variation of G.

With any of the schemes by which one has a time-varying gravitational constant, there are two kinds of times. One is the time determined by any system dominated by gravity, so-called gravitational time, such as the independent argument of an ephemeris. The other is the time determined by any system that is not dominated by gravity, so-called atomic time, such as the time kept by a hydrogen maser atomic clock. Of course, there may be more kinds of time defined for example by the strong or the weak interaction. These need not be considered here.

The use of (1), when applied to a planetary orbit, gives rise to a relation among the gravitational constant, the semi-major axis a, the period p, and the mean motion n (Counselman & Shapiro 1968),

$$2\dot{G}/G = -2\dot{a}/a = -\dot{p}/p = \dot{n}/n. \tag{2}$$

As suggested by (2), equation (1) leads to two observable effects in the solar system (Shapiro

1964a). First, the scale of the system appears to change. Second, the linearly varying period gives rise to a quadratically growing increment in the mean longitude of each body. In principle, both effects can be measured. For observations that span many orbital periods, the latter effect dominates. For Mercury, over a period of 10 years, the quadratic effect is two orders of magnitude larger than the linear.

An alternate approach is to consider explicitly the two time scales. A comparison of this with the approach represented by (1) is given by Canuto *et al.* (1983). They find that either approach will suffice for the purpose of detecting the existence of a secular variation of G. Should it be found by either approach that $\dot{G} \neq 0$, a more refined analysis would be required to distinguish the alternate formulations.

A later version of Dirac's hypothesis included two forms of matter creation (Dirac 1974). In the first, multiplicative creation, new matter appeared where old matter was present. In the second, additive creation, new matter appeared uniformly throughout the universe. These two postulated modes of matter creation give rise in turn to a pair of modifications of (2):

$$\dot{G}/G = \dot{n}/n, \tag{3}$$

$$\dot{G}/G = -\dot{n}/n. \tag{3a}$$

Again, there are the linear and quadratic observable effects. Recently, both Dirac (1979) and Canuto *et al.* (1977) have advanced theories that feature the two kinds of time. The observable consequences of these new descriptions are similar to those of (1). However, the coefficients of the linear and quadratic terms have a different ratio in the new theories.

Theoretical predictions yield $\dot{G}/G = \alpha H$ where α is a small constant and H is the Hubble constant. We will consider Mercury after nine years of observation. The accumulated effect of $\dot{G}/G = 10^{-11}$ is a mean-longitude shift such that a radar time-delay echo will show a peak increment of 8 µs. If we compare that with the uncertainty of modern radar measurement, as little as 0.2 µs, it appears that it should be very easy to determine whether or not G varies at $10^{-11} G$ per year. In fact, a simple order-of-magnitude calculation, which assumes a thousand observations and propagates the errors, yields an uncertainty of $10^{-14} G$ per year.

The problem is not so simple. Aside from the fractional solar mass loss of about 10^{-14} per year, there are problems in the analysis of the data. First, we do not know *a priori* either the initial phase or period of the Mercury orbit. For uniformly spaced and weighted data, the change from estimating one parameter to three causes the uncertainty in the estimate of $\dot{G}$ to increase by a 'masking factor' of six. But the solar-system analysis involves a very large number of parameters, not three. In a typical study of $\dot{G}$ with radar data, we find that the masking factor is about 80. Above, we used an uncertainty of measurement of 0.2 µs, about the best now available. It is not typical historically and it is almost irrelevant because of the unknown planetary topography. If we define the radar delay as a measure of the distance to the mean target-body surface, then the accuracy must include the unmodelled part of the topography; the ratio of measurement accuracy to peak effect is reduced from 40 to more like 3. Not all observations are made when the line of sight is tangent to the orbit of Mercury. In addition, there are model errors that interfere with the accuracy of the interpretation of the measurements. Our standard procedure is to perform a series of numerical experiments to uncover the effects of these errors and thus to provide realistic estimates of parameter uncertainties. When all of these factors are taken into consideration, the determination of $\dot{G}/G$ that initially appeared to have an uncertainty of 10^{-14} per year is found to have an uncertainty of 10^{-10} per year. Our published result (Reasenberg & Shapiro 1976) was a bound on $\dot{G}/G$ of $(5 \pm 10) \times 10^{-11}$ per year. Surprisingly, this result has been taken by some as evidence for a non-zero $\dot{G}$. This interpretation is, of course, not correct; we have found no significant evidence for a non-zero $\dot{G}$.

What is the prognosis for improving the accuracy of the $\dot{G}$ determination? Our present data set is listed in table 1. In some recent numerical studies we have investigated the accuracy to which we will be able to determine $\dot{G}/G$. A preliminary result is $|\dot{G}/G| < 3 \times 10^{-11}$ per year, where the bound is one standard deviation. However, by the time we finish the present series of studies, we expect to see a further reduction in the uncertainty by a factor of at least three. What are the limits to the accuracy with which $\dot{G}$ can be estimated using solar-system data? As previously noted, the uncertainty of the solar mass loss rate, about $10^{-14} M_\odot$ per year, sets a bound on the accuracy of a $\dot{G}$ estimate from planetary observations. The asteroids probably represent the most serious present limit. We now estimate the masses of eight large asteroids. Unfortunately, there are many asteroids excluded from our solutions that are nearly as large as some that are included. These and the thousands of smaller asteroids generate a 'gravitational noise' that is nearly impossible to model correctly, although various approximations are possible. This gravitational noise probably sets a bound of 0.5×10^{-11} per year to $\sigma(\dot{G}/G)$ determined from observations of the motion of Mars (J. Williams, Jet Propulsion Laboratory, personal communication 1983).

4. The fourth test of relativity

The Shapiro (1964 b) time-delay effect provides the basis for the fourth test, the most stringent solar-system test of relativity (compare with tests of the underlying principle of equivalence by means of the gravitational redshift (Vessot et al. 1980) and references therein). For a signal that passes close to the Sun during a round trip from Earth to a spacecraft or planet, the round-trip propagation time is that expected from Euclidean geometry plus an additional term,

$$\Delta\tau = \frac{2r_0}{c} S \ln\left[\frac{r_e + r_p + R}{r_e + r_p - R}\right], \tag{4}$$

where $r_0 = 3\,\text{km}$, $2r_0/c = 20\,\mu\text{s}$, and $S = \frac{1}{2}(1+\gamma)$. In (4), r_e, r_p, and R are, respectively, the distances from the Sun to the Earth, from the Sun to the target, and from the Earth to the target; γ is one of the parameters of the PPN formalism. (See Will 1981, and references therein.) For an impact parameter d substantially less than r_p or r_e:

$$\frac{r_e + r_p + R}{r_e + r_p - R} \approx \frac{4 r_e r_p}{d^2}. \tag{5}$$

In estimating γ from time-delay data, as in the other uses of the time-delay data, there are three principal problems:

(1) determination of the location of the end points of the observations;

(2) measurement of the time delay to high accuracy, including the calibration of the instrumentation; and

(3) correction of the measured time delay to the equivalent vacuum delay.

For the time delay experiment, the solar corona has traditionally been the most serious obstacle. The corona is highly variable: fluctuations are of the order of the mean. However, a simple approximate expression for the average effect of the corona on signal delay under quiet-Sun conditions can be given:

$$\tau_p = 300/f^2 d \qquad (6)$$

where τ_p is in µs, f is in GHz, and d, the impact parameter of the signal, is in solar radii.

There are several important results from pre-Viking analyses of time-delay experiments. Based on radar data through 1971, both X-band (8 GHz) and U.H.F. (400 MHz), the coefficient of the Shapiro effect, $\frac{1}{2}(1+\gamma)$, was estimated to be 1.01 with an uncertainty of 5 %. As a result of adding radar data taken in 1972, the estimate became 1.00 ± 4 %. An analysis of tracking data from the Sun-orbiting Mariners 6 and 7 yielded an estimate of 1.00 ± 3 %. Finally, there was the Mariner-9 mission: the spacecraft was placed in a 12 h orbit around Mars. Because of the short period of the orbit (cf. Mariners 6 and 7) the effects of unmodelled accelerations could not accumulate pathologically. This advantage was in part offset by the effect of the poorly known irregularities of the Mars gravitational potential. That mission yielded a measurement of the Shapiro effect of 1.00 ± 2 %.

The Viking mission provided a dramatic improvement. There were four Vikings, two orbiters and two landers, each equipped with a high-gain antenna. For the Vikings, the time delay observable far from superior conjunction has a precision of about 10 ns, that is, the distance measurements are uncertain by roughly the height of a person. Near superior conjunction, the measurement uncertainty is dwarfed by the plasma problem. Each Orbiter was equipped with a transponder that would receive an S-band signal from the ground, and send back coherent S-band and X-band signals. From the difference in the delays of these returning signals, one can determine the plasma columnar content of the path between Earth and Mars. Because they were connected to a massive body, the Landers, unlike Orbiters, were not buffeted by non-gravitational low-thrust forces. The Landers' orbits with respect to the centre of mass of Mars are relatively easy to model to high accuracy; the Landers do not execute a substantial random walk. True, there are small geophysical effects. The spin-rate of Mars is uneven and that causes the Lander to move a few metres with respect to the prediction of a simple rotation model. Mars may wobble with about a 190 day period, corresponding to the Earth's 400 day wobble. But these effects are very small compared with the kinds of effects which have plagued the analysis of spacecraft data.

Figure 1 shows the observing situation. A signal is sent from Earth, up to the Lander, and returns to the Earth which has moved since the signal was sent. A second tracking station is used to send an S-band signal to an Orbiter which returns signals at S-band and at X-band. Thus, it is the path from the Orbiter to Earth that provides the only measure of the plasma in the vicinity of the Sun between the Earth and Mars. We make the approximation that all the plasma is in a thin screen perpendicular to that line of sight and containing the Sun. In that approximation, we can determine the plasma contribution to the Lander signal if we are willing to ignore the fact that the four paths do not pierce the thin screen in the same place. The Lander plasma delay is calculated by assuming that the down-link contribution is the same as it is for the Orbiter and the up-link contribution is the same as it was for the Orbiter down-link at a time earlier by one Sun-Mars-Sun propagation time.

A preliminary analysis was done based on observations made during 40 days surrounding the first superior conjunction on 25 November 1976. This first analysis was done very crudely in comparison with our present work. Nonetheless, it yielded $\frac{1}{2}(1+\gamma) = 1 \pm 0.005$ (Shapiro et al. 1977), a factor of 4 improvement over the results of the Mariner-9 (relativity) experiment.

When we did a more formal analysis based on 14 months of observation, we obtained a much more accurate result. By then, we had developed a systematic procedure for correcting the effect of the plasma. All aspects of the analysis were under computer control, with no potentially prejudicial hand intervention. Most of the residuals are less than 100 ns; they get worse near superior conjunction as is to be expected. From this analysis we have determined that the data are consistent with relativity: $\frac{1}{2}(1+\gamma) = 1.000 \pm 0.001$ (Reasenberg et al. 1979). In terms of the Brans–Dicke theory, this result implies that ω is greater than 500. A histogram of the distribution of the residuals shows that, as is often the case with real data, the distribution resembles a Gaussian except that the tails are too high. This and other evidence of systematic error was taken into consideration in determining the uncertainty.

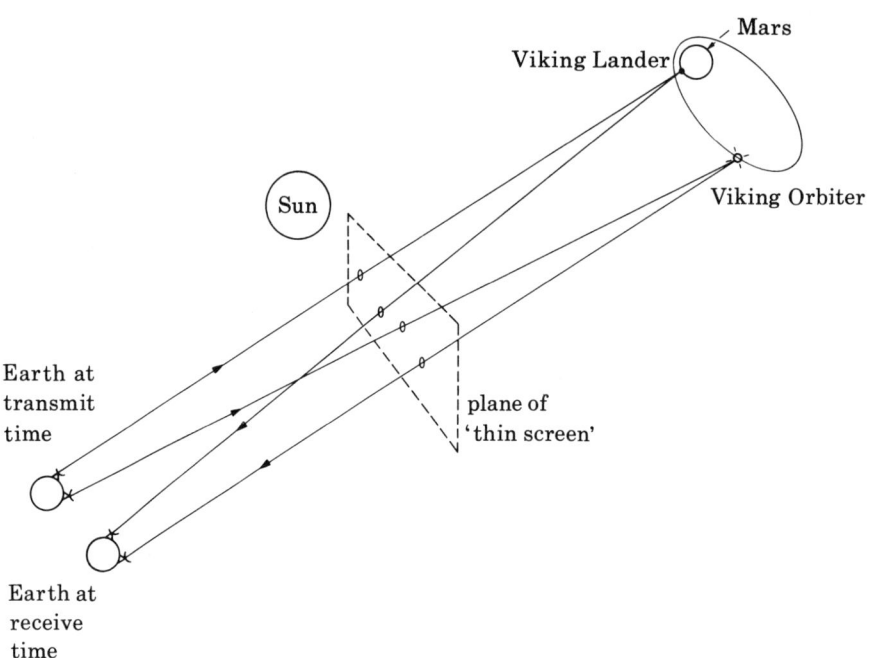

FIGURE 1. The observing geometry, not to scale.

We are not yet finished with the analysis of this experiment. What are the prognoses for further improvement? There have been two superior conjunctions for which the equipment was working. The above results are based on the first of those. We had four objects, all of which have now failed after functioning far past the end of their design life. The second Orbiter failed on 5 August 1980, after which there was no way of measuring the contribution to the delay from the plasma between Earth and Mars. Lander One, which was the last Viking to fail, was working until late 1982; during the last several months, useful data were being obtained about twice per month. Although there is no way to correct these data for the effect of the plasma, far from superior conjunction the plasma effects are small and relatively stable; these data are quite useful. They are directly applicable to improving the planetary ephemerides and thus indirectly useful for the relativity tests.

We have done some numerical experiments to determine what sort of accuracies we ought to expect from a complete analysis of all of the data now available. These suggest that we can decrease the uncertainty in the estimate of $\frac{1}{2}(1+\gamma)$ by a factor of two or three below our best published results. It will not be until we finish the present series of studies, that we will be able to make a new and reliable determination of the time-delay coefficient.

5. POINTS

The results of all accepted experiments are consistent with the predictions of general relativity. However, with the exception of the binary pulsar study (Taylor & Weisberg 1982), all the experiments depend on first-order post-Newtonian effects. One must wonder whether it is particularly fruitful at this time to invest in further improvement of first-order tests. Does it seem reasonable that general relativity fails deep in first order or should we be looking elsewhere? I believe that it is no longer reasonable to make substantial investments in small improvements to first-order tests of the kinds already performed. New kinds of first-order tests, such as those based on motion of the source (e.g. the Stanford gyroscope experiment), remain interesting.

How could one perform a second-order test of relativity? The problem is difficult because the solar potential at the limb of the Sun is 2×10^{-6}. A second-order experiment must be at least three orders of magnitude more sensitive than the present solar-system tests: a considerable challenge.

A second-order test, if it is to be done within the solar system, will likely utilize the potential close to the limb of the Sun. Putting an instrument package close to the Sun causes substantial engineering problems. Such a mission, STARPROBE (Reasenberg *et al.* 1982), is being considered by NASA but is now on hold and its future is uncertain. Thus we must find a way of doing an experiment in which we use photons that nearly graze the Sun. The two reasonable possibilities, the light deflexion experiment and the time-delay experiment, depend on essentially the same phenomenon. For the foreseeable future, the time-delay experiment ineluctably involves spacecraft tracking and the NASA deep space network which, for a combination of technical and political reasons, seems unlikely in the foreseeable future to be able to yield data with the accuracy necessary to measure the second-order time delay. Thus we are limited to considering a light deflexion experiment.

At the solar limb, the second-order light deflexion is 10.9 µs (Epstein & Shapiro 1980; Fischbach & Freeman 1980; Richter & Matzner 1983), five orders smaller than the 1.75 s first-order deflexion. The plasma-related problems, which are a limiting factor in first-order experiments using microwave signals, demand the use of substantially higher frequency signals for a second-order test. A reasonable solution is to use signals in the optical region. Since the resolution of any angle-determining instrument is closely related to the ratio of the wavelength to the size of the instrument, the optical region offers an additional advantage. Thus we are led to the possibility of using optical interferometry as a means of determining the light deflexion to second order in the solar potential. Any device that would perform such an experiment would have to be above Earth's atmosphere because of the large corrupting effect of atmospheric refraction.

Figure 2 shows a conceptual design of an instrument to perform the second-order deflexion experiment by means of Precision Optical INTerferometry in Space (POINTS). It is a dual optical interferometer with nominal baselines of 10 m. There are four one-metre telescopes that bring together light at the two central sections, where fringes are formed and analysed. A beam splitter, lens, and prism yield a spectrum of the interference (i.e. a channelled spectrum) which is allowed to fall on a linear array of photon-counting detectors. The complementary channelled spectra from the two exit ports of the beam splitter contain all the information necessary to determine the displacement of the source from the optical axis of the interferometer. As the principal axis of such an instrument is moved away from the source, the number of channels in the spectrum increases. The ability of the linear detector array to separate the undulations in the channelled spectrum sets the limit to the allowed mispointing of the instrument. In the nominal design the linear array has 1000 detectors; the pointing error limit is $\pm 2.5''$.

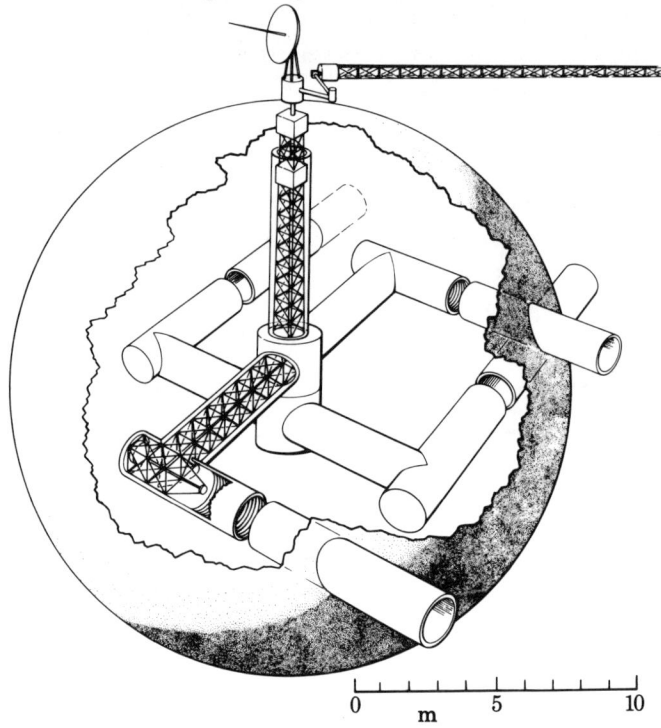

FIGURE 2. An artist's rendition of the proposed astrometric interferometer, POINTS. The satellite diameter is 20 m; each of the two interferometers has a pair of 1 m telescopes at the ends of a 10 m baseline. This design is the result of a collaboration with the C. S. Draper Laboratory.

A simple photon-statistics analysis of the system indicates that a single interferometer observing a 10th magnitude source at 7000 K has a measurement uncertainty of 1 µs with an integration of time of 150 s. Thus when POINTS is used to observe bright stars, it is a fast measuring device. There would be time to conduct the relativity experiment along with several other kinds of research: looking for planetary systems, determining the lower rungs of the cosmic distance ladder by measuring the distance to Cepheid variables, and numerous other astrophysically interesting studies. The diversity of application is important, not only to the astrophysicist directly, but also indirectly. It is the speed of measurement and the potential diversity of application that will make possible a large constituency from the astronomical community to back the development of POINTS.

The spherical enclosure is intended to reduce severe thermal problems such an instrument would experience if exposed directly to the Sun. Several of the technical aspects of this cartoon are now obsolete, as we have come to understand the instrument better, but the concept is still considered valid. The two interferometers shown are at right angles. This configuration maximizes the number of reference stars available for a given target star and simplifies the measurement of the angle between the optical axes of the two interferometers.

To achieve the required accuracy with POINTS, the locations of the optical components must be known with an uncertainty of no more than about 0.1 Å,† 10^{-12} of the baseline length of 10 m. This cannot be done entirely by the use of stable materials. In principle, the required measurements can be performed with onboard laser metrology. However, consider what it means to know the location of an extended object like a telescope mirror to 0.1 Å. Such an object is a chunk of glass, quartz, or some composite material which has a non-zero thermal expansion coefficient, distorts under mechanical stress, and may creep. Optical components will not only change temperature, but they will have varying temperature distributions and the corresponding distortions. Thus, it will not be sufficient to keep track of the locations of a few reference points on a large optical element; the entire surface of the element must be monitored. Fortunately, we know in principle how to construct a system to provide the required 'full aperture' metrology.

The problem now is one of detailed engineering and forming a constituency. It seems unlikely that we can start construction of POINTS during this decade. Perhaps some advanced design work can be started by the end of the decade, and the device can be constructed in the next decade for launch either in the late 1990s or the early 21st century.

6. Conclusion

There is now a small, international industry dedicated to testing relativistic theories of gravitation. In the decades since the emergence of general relativity, and particularly in the last 20 years, considerable experimental progress has been made. There are several successful first-order tests and the strong evidence from PSR 1916+13, all supporting the validity of general relativity. We appear now to be on the threshold of a new era. It is time to investigate additional phenomena including the Lense–Thirring effect, second-order effects, and the interaction of gravitation with quantum mechanical systems. These are among the experimental challenges for the coming decades.

I thank N. L. Murphy for her careful preparation of the text. I am grateful to I. I. Shapiro for his comments. This work was supported in part by the Smithsonian Astrophysical Observatory, the Secretary's Fluid Research Fund of the Smithsonian Institution, the National Aeronautics and Space Administration (NASA) through the Mars Data Analysis Program Contract NSG-7556, and the National Science Foundation through grant PHY-8106036.

References

Canuto, V., Adams, P. J., Hsieh, S.-H. & Tsiang, E. 1977 *Phys. Rev.* D **16**, 1643–1663.
Canuto, V. M., Goldman, I. & Shapiro, I. I. Testing the strong equivalence principle by radio ranging. (In preparation.)
Counselman, C. C., Kent, S. M., Knight, C. A., Shapiro, I. I., Clark, T. A., Hinteregger, H. F., Rogers, A. E. E. & Whitney A. R. 1974 *Phys. Rev. Lett.* **33**, 1621.
Counselman, C. C. & Shapiro, I. I. 1968 *Science, N.Y.* **162**, 352–355.
Dirac, P. A. M. 1937 *Nature, Lond.* **139**, 323.
Dirac, P. A. M. 1938 *Proc. R. Soc. Lond.* A **165**, 199.
Dirac, P. A. M. 1974 *Proc. R. Soc. Lond.* A **338**, 439–446.
Dirac, P. A. M. 1979 *Proc. R. Soc. Lond.* A **365**, 19–30.
Dyson, F. J. 1972 In *Aspects of quantum theory* (ed. A. Salam & E. P. Wigner), pp. 213–236. Cambridge University Press.

† 1 Å = 10^{-10} m.

Epstein, R. & Shapiro, I. I. 1980 *Phys. Rev.* D **22**, 2947–2949.
Fischbach, E. & Freeman, B. S. 1980 *Phys. Rev.* D **22**, 2950–2952.
Fomalont, E. B. & Sramek, R. A. 1977 *Comments Astrophys.* **7**, 19.
Reasenberg, R. D. 1980 In *Cosmology and Gravitation* (ed. P. G. Bergmann and V. DeSabbata). New York and London: Plenum Press.
Reasenberg, R. D. 1983 In *Proceedings of the Third Marcel Grossmann Meeting on the Recent Developments of General Relativity* (In the press.)
Reasenberg, R. D., Anderson, J. D., DeBra, D. B., Shapiro, I. I., Ulrich, R. K. & Vessot, R. F. C. 1982 In *STARPROBE scientific rationale, a report of the ad hoc working groups* (ed. J. H. Underwood & J. E. Randolph). JPL Publication 82–49, Pasadena.
Reasenberg, R. D. & Shapiro, I. I. 1976 In *Atomic masses and fundamental constants*, vol. 5 (ed. J. H. Sanders & A. H. Wapstra), pp. 643–646. New York and London: Plenum Press.
Reasenberg, R. D., Shapiro, I. I., MacNeil, P. E., Goldstein, R. B., Breidenthal, J. C., Brenkle, J. P., Cain, D. L., Kaufman, T. M., Komarek, T. A. & Zygielbaum, A. I. 1979 *Astrophys. J. Lett.* **234**, 219.
Richter, G. W. & Matzner, R. A. 1983 *Phys. Rev.* D. (In the press.)
Shapiro, I. I. 1964a *Effects of general relativity on interplanetary time-delay measurements*. M.I.T. Lincoln Laboratory technical report 368, Lexington, Massachusetts.
Shapiro, I. I. 1964b *Phys. Rev. Lett.* **13**, 789.
Shapiro, I. I., Counselman, C. C. & King, R. W. 1976 *Phys. Rev. Lett.* **36**, 555.
Shapiro, I. I., Reasenberg, R. D., MacNeil, P. E., Goldstein, R. B., Brenkle, J. P., Cain, D. L., Komarek, T., Zygielbaum, A. I., Cuddihy, W. F. & Michael, Jr, W. H. 1977 *J. geophys. Res.* **82**, 4329.
Shapiro, I. I., Smith, W. B., Ash, M. E., Ingalls, R. P. & Pettengill, G. H. 1971 *Phys. Rev. Lett.* **26**, 27–30.
Taylor, J. H. & Weisberg, J. M. 1982 *Astrophys. J.* **253**, 908.
Texas Mauritanian Eclipse Team 1976 *Astr. J.* **81**, 452.
Vessot, R. F. C., Levine, M. W., Mattison, E. M., Blomberg, E. L., Hoffman, T. E., Nystrom, G. U., Farrel, B. F., Decher, R., Eby, P. B., Baugher, C. R., Watts, J. W., Teuber, D. L. & Wills, F. D. 1980 *Phys. Rev. Lett.* **45**, 2081.
Will, C. M. 1981 *Theory and experiment in gravitational physics*. Cambridge: Cambridge University Press.

Limits on the variability of coupling constants from the Oklo natural reactor

By J. M. Irvine

Schuster Laboratory, Department of Theoretical Physics, University of Manchester, Oxford Road, Manchester M13 9PL, U.K.

Thermal neutron capture cross sections in fission fragments need to be known to great precision in the nuclear power industry. These cross sections are frequently dominated by extremely narrow neutron capture resonances. An analysis of isotopic abundances at the site of a prehistoric natural reactor at Oklo in West Africa suggests that such a resonance in samarium-149 has moved by less than 0.01 eV in the past 2×10^9 years and this is used to place limits on the variability of coupling constants over this period.

1. Introduction

Typical nuclear interaction energies are measured in mega electron volts while thermal neutron capture cross sections in heavy nuclei are frequently dominated by extremely narrow resonances known to a precision of milli electron volts. These resonances thus provide a one part in 10^9 probe of the single-particle couplings in nuclei.

The discovery of the site of a prehistoric natural nuclear reactor at Oklo in Gabon, West Africa which was critical some 2×10^9 years ago provides us with geological samples, the isotopic abundances in which can be used to extract information on the operational characteristics of the reactor. Among these are thermal neutron capture cross sections. Limits on the variability of these cross sections place restrictions on the variability of the energies of single-neutron capture-resonances from which we may deduce limits on the variability of particle coupling strengths.

2. Reactor physics

The primary source of energy in a thermal nuclear reactor is the induced neutron fission of ^{235}U which is accompanied by the release of 2-3 neutrons,

$$\text{n} + {}^{235}\text{U} \rightarrow \text{fission fragments} + \text{2-3 neutrons}. \tag{1}$$

When conditions are right for some of these neutrons released in fission to induce further fission then we have the possibility of a self-sustaining chain reaction.

The first problem that we encounter is that the neutrons released in fission have energies typically in the range 1–2 MeV while the neutron's effectiveness in inducing fission increases with the length of time it spends in the nucleus, that is, the lower its velocity and hence its energy. Neutrons in equilibrium with their surroundings will have a thermal distribution and in a reactor core at a temperature of $T \approx 1000$ K; this implies a typical energy $kT \approx 86$ meV. There is thus the need for a moderator to slow down the fission neutrons to thermal energies where they can effectively induce further fission. The most effective moderators are protons, and in the form of hydrogen these are readily available in water. A slight concern is that, besides slowing down neutrons, protons have a thermal neutron capture cross section of 0.33×10^{-24} cm^2 which removes neutrons from the system so that they are no longer available to induce fission in the ^{235}U. This can be overcome by using deuterons in heavy water, rather than protons, as a moderator.

The second problem is the low relative abundance of ^{235}U which reduces the probability that a thermal neutron will encounter such a nucleus before it leaves the system or is captured by some other nucleus. The world-average ^{235}U isotopic abundance is $0.7202 \pm 0.00014\,\%$, the remainder being ^{238}U which is not amenable to thermal neutron induced fission. To overcome this problem uranium fuel for man-made reactors is artificially enriched, that is, the relative abundance of ^{235}U is increased. Clearly a light-water moderated reactor requires a higher level of enrichment than does one moderated by heavy water.

3. Fossil reactors
(a) General considerations

The relative shortage of ^{235}U can be traced to the half-lives of the uranium isotopes which are 7×10^8 years and 5×10^9 years for ^{235}U and ^{238}U respectively. Thus in the past there was more uranium than there is today and a higher proportion of it was ^{235}U. The half-lives of all other uranium isotopes are so much shorter that their natural abundances have been negligible throughout geological times. In table 1 we present the natural abundance of ^{235}U in the past based upon the present day level of 0.72 %. For comparison the degree of uranium enrichment required by man-made light-water reactors is 3–5 %. Thus we can see that it is extremely unlikely that an accidental accumulation of uranium-rich ore and water could have gone critical in the last 1.5×10^9 years.

TABLE 1. THE PAST ENRICHMENT OF URANIUM ORES

age ($10^6 \times$ years) ...	0	700	1400	2100	2800
percentage ^{235}U	0.72	1.3	2.3	4.0	7.0

Besides the necessary level of enrichment, an additional requirement before a natural nuclear reactor could go critical would be the concentration of a sufficient mass of uranium. Impurities can migrate through crystalline materials ending up concentrated along fault lines and planes producing thin veins of mineral deposits. The chances of an essentially one- or two-dimensional reactor core achieving criticality are negligible. The development of a truly three-dimensional ore rich abundance requires mass transport and deposition of the element concerned. This is most easily achieved if the material is water soluble. In the case of uranium this means the existence of uranium oxides. It is generally believed that atmospheric oxygen first became available through photosynthesis some 2×10^9 years ago. Thus conditions suitable to the formation of a natural nuclear reactor seem limited to a period 1.5–2×10^9 years ago.

(b) Oklo

In 1972 the French C.E.A., monitoring uranium ores for their nuclear fuel enrichment programme, discovered samples that were heavily depleted in ^{235}U. Samples as low as 0.2 % were found and those with 0.4 % ^{235}U were typical. These ores came from the Oklo mines in Gabon, West Africa. A study of the site has led to the interpretation that the ^{235}U depletion is due to burn up in a natural reactor core.

The present-day characteristics of the Oklo ore bodies are consistent with conditions 1.8×10^9 years ago being similar to those in a man-made p.w.r. They suggest that criticality was achieved

$1.84 \pm 0.07 \times 10^9$ years ago, that the reactor was operational for $2.29 \pm 0.7 \times 10^5$ years (although whether the operation was pulsed or continuous is not clear) and that integrated neutron fluxes of $1-2 \times 10^{21}$ neutrons cm^{-2} were experienced.

In support of these conclusions we present an analysis of neodymium isotopic abundances in table 2. In columns 1 and 2 of this table we show the isotopic abundances of two samples from Oklo. While there is generally a good correlation between the two samples there is an exception in the case of ^{142}Nd which is almost four times more prevalent in sample 2. Neither of the Oklo samples show any similarity to the world-average abundances for these isotopes presented in column 3. The only one of these isotopes not produced as a fission product is ^{142}Nd. We can use the ^{142}Nd abundances and the 'natural' abundances from column 3 to obtain the 'abnormal' abundances for the Oklo samples given in columns 4 and 5. The two samples now show an enhanced correlation and a striking resemblance to the ^{235}U fission yields presented in column 6.

TABLE 2. NEODYMIUM ABUNDANCES (%)

	Oklo 1	Oklo 2	world average	Oklo 1*	Oklo 2*	^{235}U$_f$
^{142}Nd	1.38	5.49	27.11	0	0	0
^{143}Nd	22.1	23.0	12.17	22.6	25.7	28.8
^{144}Nd	32.0	28.2	23.85	32.4	29.3	26.5
^{145}Nd	17.5	16.3	8.30	18.05	18.4	18.9
^{146}Nd	15.6	15.4	17.22	15.55	14.9	14.4
^{148}Nd	8.01	7.70	5.73	8.13	8.20	8.26
^{150}Nd	3.40	3.90	5.26	3.28	3.46	3.12

The only discrepancy between the samples and the fission yields is the slightly low readings for the odd isotopes ^{143}Nd, ^{145}Nd and the slightly high readings for the even isotopes ^{144}Nd, ^{146}Nd in the samples. This is to be expected if the ore has been exposed to a neutron flux in the reactor core, since the odd neodymium isotopes have substantial thermal neutron capture cross sections for the formation of the even neodymium isotopes. This is clarified in table 3 where we have added together the ^{143}Nd and ^{144}Nd and the ^{145}Nd and ^{146}Nd abundances to yield an enhanced agreement between the averaged Oklo 'abnormal' abundances and those resulting from ^{235}U fission. The relative ^{143}Nd to ^{144}Nd and ^{145}Nd to ^{146}Nd abundances are consistent with exposure to an integrated thermal neutron flux of 10^{21} neutrons cm^{-2}.

TABLE 3. NEODYMIUM ABUNDANCES (%)

	143+144	145+146	148	150
world average	36.02	25.52	5.73	5.62
^{235}U$_f$	55.18	33.53	8.16	3.13
Oklo	54.95	33.49	8.25	3.34
95% U$_f$ + 5% Pu$_f$	54.78	33.65	8.28	3.29

In any nuclear reactor some fast neutrons will be captured by ^{238}U nuclei before they are moderated. The resulting ^{239}U can beta-decay to form ^{239}Pu

$$n + {}^{238}\text{U} \to {}^{239}\text{U} \underset{\beta^-}{\to} {}^{239}\text{Np} \underset{\beta^-}{\to} {}^{239}\text{Pu}, \qquad (2)$$

which is susceptible to thermal neutron induced fission. The fission yields from ^{239}Pu are slightly different from those of ^{235}U. A best fit to the Oklo abundances is given by a 5% contamination by ^{239}Pu fission yields (see the bottom row of table 3). The relative plutonium to uranium fission yields is a sensitive measure of the moderator to uranium content of the reactor core and suggests for the Oklo reactor a water to uranium abundance of 15 %.

(c) *The samarium isotopes*

In table 4 we present an analysis of the isotopic abundances for the samarium isotopes. The results are reminiscent of those for the neodymium isotopes. The Oklo abundances are very different from the world-averages; ^{144}Sm is not a fission product and can be used to determine the 'abnormal' Oklo samarium abundances; The abnormal samarium abundances are reminiscent of the ^{235}U fission yields; In the Oklo samples the odd isotopes show a depletion relative to the even isotopes consistent with an integrated neutron flux of 10^{21} neutrons cm^{-2}. The difference between the samarium and neodymium analysis is the extreme depletion of ^{149}Sm due to enormous thermal neutron capture cross section of 4.2×10^{-20} cm^2. This can be compared with the thermal neutron capture cross sections for ^{143}Nd, ^{145}Nd and ^{147}Sm which are 3.3×10^{-22} cm^2, 5×10^{-24} cm^2 and 9×10^{-23} cm^2 respectively.

TABLE 4. SAMARIUM ABUNDANCES (%)

	Oklo	world average		Oklo*	^{235}U$_f$
^{144}Sm	1.2	3.2		0	0
^{147}Sm	40.0	15.0	26.3	61.2	61.6
^{148}Sm	5.3	11.3			
^{149}Sm	3.9	13.9	31.4	27.3	29.3
^{150}Sm	21.4	17.5			
^{152}Sm	17.5	26.7		10.5	7.2
^{154}Sm	10.7	22.5		1.8	1.9

With the integrated neutron flux fixed at 10^{21} neutrons cm^{-2} the ^{149}Sm/^{147}Sm abundance ratio requires that the ^{149}Sm thermal neutron capture cross section during the operation of the Oklo reactor must have been within 10 % of its present day value.

Details of the findings at the Oklo mines are presented in an I.A.E.A. 1975 symposium proceedings.

4. CONCLUSIONS

The extraordinarily large ^{149}Sm thermal neutron capture cross section is due to the existence of a well established neutron resonance at 98 meV of width 63 meV which is firmly spanned by the thermal neutron distribution. In contrast the corresponding resonance in ^{147}Sm is at 3.4 eV and is well outside the thermal régime.

The total thermal neutron capture cross section in ^{149}Sm is the result of enfolding the energy dependent neutron capture cross section with the thermal distribution function. The thermal distribution is peaked at $\frac{1}{2}kT$ and has a width of order kT (remember $T \approx 1000$ K implies $kT \approx 86$ meV) whereas the neutron capture cross section is dominated by an extremely narrow resonance of width 63 meV. The total thermal neutron capture cross section is thus extremely sensitive to the relative position of these two highly peaked functions. Today's value of 4.2×10^{-20} cm^2 is a measure of the proximity of the resonance at 98 meV to the thermal peak. In order that the cross section was within 10 % of today's value during the operation of the Oklo reactor this neutron resonance must have been within 0.01 eV of its present day position.

The interaction of a neutron with a rare earth nucleus like ^{149}Sm is well described by an optical potential *ca.* 50 MeV deep and hence the limitation on the variation of the neutron resonance to less than 0.01 eV translates into a limit on the variation of the neutron coupling to the nucleus of 2×10^{-10} over the past 2×10^9 years or less than one part in 10^{19} per year over this period. Since this coupling comes *ca.* 95 % from the strong interactions, *ca.* 5 % from electromagnetic effects and *ca.* 10^{-5} % from the weak interactions, we deduce that the corresponding coupling constants have altered by less than one part in 10^{19}, 5×10^{17} and 10^{12} per year respectively over the past 2×10^9 years.

REFERENCE

The Oklo Phenomenon 1975 I.A.E.A. Symposium proceedings. Vienna: I.A.E.I.

Implications of quasar spectroscopy for constancy of constants

By B. E. J. PAGEL

Royal Greenwich Observatory, Herstmonceux Castle, Hailsham, East Sussex BN27 1RP, U.K.

Given that the redshifts of quasars are due to the universal expansion, the details of their spectra can be used to check the constancy of certain dimensionless ratios of physics over look-back times comparable with the Hubble time and distances exceeding that of the particle horizon in the Einstein–de Sitter universe. With $\alpha, g_p, g_e, m_p, m_e$ denoting the fine structure constant, and the gyromagnetic ratios and masses of protons and electrons respectively, the following upper limits to variability over such times and distances have been derived in this way:

effect	quantity	approximate 3σ upper limit to variation
optical doublet splittings	α	3%
comparison of optical and 21 cm redshifts	$\alpha^2 g_p m_e/m_p$ (or $\alpha^2(g_p/g_e)(m_e/m_p)$)	10^{-3}
comparison of hydrogen and metal redshifts	m_e/m_p	50%

1. INTRODUCTION

The purpose of this short paper is to review the contribution made by astrophysical observations of the spectra of distant galaxies and quasars to the placing of limits on the past variability of certain dimensionless ratios of physics: specifically, the fine structure constant α, the electron: proton mass ratio $m_e:m_p$ and the gyromagnetic ratio g_p of the proton (or more strictly its ratio to the corresponding value g_e of the electron), all of which enter the expression for the energy level shift due to hyperfine structure. The limits on possible variability are not as impressive as those derived from certain terrestrial measurements, in particular those of the Oklo phenomenon (Shlyakhter 1976; Irvine 1983), but in compensation they take us over large look-back times corresponding to redshifts z of up to 2.5 or more (compared with $z \approx 0.4$ for the epoch of origin of the solar system and $z \approx 0.15$ for Oklo); furthermore, they enable us to compare parts of the universe that may be causally unconnected with one another, and from a philosophical viewpoint this could be the most interesting part.

2. CONSTANCY IN TIME

While there is, by now, a wealth of literature on quasars that could be used to carry out elaborate statistical analyses if desired, there have in fact been only a few investigations specifically directed to testing for variability of atomic constants, presumably because interest has tended to be damped down by the consistently negative results of such experiments (cf. Dyson 1972; Yahil 1975). In the 1960s Bahcall and collaborators (Bahcall & Salpeter 1964; Bahcall et al. 1967;

Bahcall & Schmidt 1967), in response to a suggestion by Gamow that α (representing the electronic charge) might increase proportionally to time in such a way as to satisfy Dirac's 'large numbers' hypothesis, showed that this idea was ruled out by the existence of unchanged relative doublet splittings of optical emission lines due to [O III] and [Ne III] in radio galaxies with $z \approx 0.2$ and of ultraviolet (in the rest frame) absorption lines of Si II and Si IV in the QSO 3C 191 with $z = 1.95$. (QSO emission lines are unsuitable for this game because they are too broad.) These and many other more recent data (especially for C IV $\lambda1548.20$ and 1550.77) indicate that α is indeed constant with a standard error of about 1 % for redshifts up to 2.5 or so, i.e. look-back times between 0.7 and 0.85 of the age of the Universe if $0 < \Omega < 1$ and $\Lambda = 0$.

Ten years later, Wolfe et al. (1976) studied the BL Lac object AO 0235+164 and discovered a 21 cm absorption line of HI having precisely the same redshift of about 0.5 as the ultraviolet doublet of Mg$^+$ previously identified in the optical spectrum by Burbidge et al. (1976). Since the hyperfine splitting (in Rydbergs) is proportional to

$$\alpha^2(g_p/g_e)(m_e/m_p) \quad \text{or} \quad \tfrac{1}{2}\alpha^2 g_p(m_e/m_p),$$

their results place a tight limit on the past variability of this product with a 1σ uncertainty of 3×10^{-4} of itself in a Hubble time. Soon afterwards, I undertook an analysis of the component m_e/m_p in that product, by looking for a differential mass shift between hydrogen and metal lines with redshifts between 2.1 and 2.7. The ratio $m_e:m_p$ was found to be the same as now with a standard error of about 10 % (Pagel 1977) which, when combined with the previous result for hyperfine splitting, places a similar limit on the gyromagnetic ratio g_p of the proton and on the strong interaction effects that bring it about, assuming $g_e = 2$ always.

3. Constancy in causally unconnected parts of the Universe

In an extension of their 21 cm redshift studies to three other sources, Tubbs & Wolfe (1980) pointed out the significance of making observations of highly redshifted objects in different directions in the sky. Such objects can either be causally disjoint from one another, in that their co-moving separation exceeds the radius of the past light cone at all times since the big bang, or at least be so far apart that any one source is in causal contact with large tracts of space–time that are themselves disjoint from other sources. They illustrated this situation with a diagram valid in an Einstein–de Sitter universe and showed that for all reasonable values of Ω their sample of four objects is a case of the latter type. It follows that large variations in the hyperfine splitting within the radius of any one horizon are ruled out to the extent that this quantity is affected by its value at other places within the quasar's past light cone; but if it is not so affected, then spatial variations are ruled out only in a much smaller volume bounded by the quasars themselves.

If, however, we leave the case of the 21 cm line and content ourselves with the lesser precision associated with α and m_e/m_p, then we can discuss absorption-line systems of such high redshift that we could have a case of the first kind in which the objects themselves (really the clouds producing the absorption lines) are causally disjoint from one another, because their separation θ exceeds the critical separation θ_0 given by

$$2\sin\tfrac{1}{2}\theta_0 = [(1+z)^{\frac{1}{2}} - 1]^{-1}; \quad z > 1.25,$$

in the Einstein–de Sitter universe (J. D. Barrow, personal communication). Examples taken from Pagel (1977) and elsewhere are listed in table 1 and many more could be added to these

Table 1. Selected quasars having absorption line systems with Lyman α and CIV

quasar	z_{abs}	θ_0/deg	θ/deg from 1623	reference†
PHL 957 (0100+13)	2.31	75.2 ⎫		
	2.66	66.4 ⎬	127	1
	1.72	100.7 ⎭		
0551−37	1.96	87.9	159	2
0736−06	1.91	90.2 ⎫	123	2, 3
	1.93	89.3 ⎭		
1623+27	2.05	84.1 ⎫		
	2.09	82.6 ⎬	—	4
	2.24	77.4 ⎭		

† 1, Coleman et al. (1976); 2, Young et al. (1982); 3, Carswell et al. (1976); 4, Sargent (1982).

which from CIV splittings and hydrogen-metal line concordance satisfy the constraints on variability already given. The situation is quite analogous to that of the uniformity of causally disjoint sources of the microwave background (see, for example, Tubbs & Wolfe 1980; Rees 1982) and it suggests (if $\Omega_0 \geqslant 0.5$ or so) that the constants had their present values impressed on them at a very early stage before the present expansion phase of the Universe.

I thank Dr John D. Barrow for helpful and stimulating discussions.

References

Bahcall, J. N. & Salpeter, E. E. 1965 *Astrophys. J.* **142**, 1677.
Bahcall, J. N., Sargent, W. L. W. & Schmidt, M. 1967 *Astrophys. J. Lett.* **149**, 11.
Bahcall, J. N. & Schmidt, M. 1967 *Phys. Rev. Lett.* **19**, 1294.
Boksenberg, A. & Sargent, W. L. W. 1975 *Astrophys. J.* **198**, 31.
Burbidge, E. M., Caldwell, R. D., Smith, H. E., Liebert, L. & Spinrad, H. 1976 *Astrophys. J. Lett.* **205**, 11.
Coleman, G., Carswell, R. F., Strittmatter, P. A. & Williams, R. E. 1976 *Astrophys. J.* **207**, 1.
Dyson, F. J. 1972 In *Aspects of quantum theory* (ed. A. Salam & E. P. Wigner), p. 213. Cambridge University Press.
Irvine, J. M. 1983 *Phil. Trans. R. Soc. Lond.* A **310**, 239.
Pagel, B. E. J. 1977 *Mon. Not. R. astr. Soc.* **179**, 81P.
Rees, M. J. 1982 *Phil. Trans. R. Soc. Lond.* A **307**, 97.
Sargent, W. L. W., Young, P. & Schneider, D. P. 1982 *Astrophys. J.* **256**, 374.
Shlyakhter, A. I. 1976 *Nature, Lond.* **264**, 340.
Tubbs, A. D. & Wolfe, A. M. 1980 *Astrophys. J. Lett.* **236**, 105.
Wolfe, A. M., Brown, K. L. & Roberts, M. S. 1976 *Phys. Rev. Lett.* **37**, 179.
Yahil, A. 1975 In *The interaction between science and philosophy* (ed. Y. Elkana). Atlantic Highlands, N.J.: Humanities Press.
Young, P., Sargent, W. L. W. & Boksenberg, A. 1982 *Astrophys. J. supp.* **48**, 455.

Overview of theoretical prospects for understanding the values of fundamental constants

By S. Weinberg, For.Mem.R.S.

Theory Group, Department of Physics, University of Texas, Austin, TX 78712, U.S.A.

This is a brief summary of the talk given at the Meeting.

The membership of a list of 'fundamental' constants necessarily depends on who is compiling the list. A hydrodynamicist might reasonably include the density and viscosity of water, while an atomic physicist would doubtless include the proton mass and electronic charge. This talk deals with a different sort of list: a list of the constants that appear in the laws of nature at the deepest level that we yet understand, constants whose value we cannot calculate with precision in terms of more fundamental constants, not just because the calculation is too complicated (as for the viscosity of water or the mass of the proton) but because we do not know of anything more fundamental. The membership of such a list of fundamental constants thus reflects our present understanding of fundamental physics. Also, each constant on the list is a challenge for future work, to try to explain its value.

The parameters that appear at the most fundamental level in our present theories of elementary particles are (1) the electroweak and strong gauge couplings g_1, g_2, g_3, and (2) the masses and self couplings of the 'Higgs' scalars, and (3) the coupling constants for the interaction of the scalars to quarks and leptons. The gauge couplings themselves determine† the observables $e^2 = g_1^2 g_2^2/(g_1^2 + g_2^2)$ and $\sin^2\theta = g_1^2/(g_1^2 + g_2^2)$, which are experimentally known to be $4\pi/137$ and 0.22. The scalar masses and self-couplings determine the scalar vacuum expectation values, and hence the Fermi coupling $G_F = \langle\phi^0\rangle^{-2}/\sqrt{2}$, which is experimentally known to be $(293\,\text{GeV})^{-2}$. Combined with the gauge couplings, these vacuum expectation values also determine the W and Z masses, with predicted values of about 83 and 93 GeV, which are now happily in agreement with experiment. Finally, the scalar vacuum expectation values and scalar–fermion couplings determine the quark and lepton masses, which experimentally range over more than three orders of magnitude. From these experimental masses we deduce that the scalar–fermion coupling constants are much smaller than the gauge couplings and very different from each other, but we have no clear idea of why this should be so.

It is somewhat misleading to list the gauge couplings as fundamental parameters. We know in the case of quantum chromodynamics that the QCD coupling is not a constant; rather g_3^2 varies with the energy E as $24\pi^2/25 \ln(E/\Lambda_3)$, (for E below the bottom mass). The constant Λ_3 determines the general scale of strong interaction physics, including the current-algebra constant F_π and the 'constituent quark' mass m_q. Experimentally $\Lambda_3 \sim 150\,\text{MeV}$, $F_\pi \sim 190\,\text{MeV}$, and $m_q \approx \frac{1}{3} m_N = 310\,\text{MeV}$. We do not yet know how to calculate F_π and m_q from Λ_3. The only place

---

† When I say that one set of constants 'determines' a second set, I mean that we think that the first set is more fundamental, and that the values of the second set are what they are because of the values taken by the first set, in the way for instance that the energy levels of atoms are what they are because of the values taken by the electron's mass and charge and Planck's constant. I do not mean that we have historically deduced the values of the second set from those of the first set: in fact the opposite is more likely to be the case.

in low energy strong interaction physics where another scale factor intrudes is in the pion mass, which is given in terms of the 'bare quark' mass $m_{q'}$ (the one determined by the scalar couplings) by $m_\pi^2 \approx \Lambda_3 m_{q'}$; it is from this relation that we deduce the bare u and d quark mass scale of a few MeV.

It is possible that the scalars are not really elementary, but bound by some sort of extra strong or 'technicolour' force, with a scale parameter Λ_4 very different from Λ_3. In this case it is Λ_4 that determines the Fermi coupling, from whose value we deduce that Λ_4 must be about 300 GeV. The problem with this view is that it is difficult to see where the quark and lepton masses come from. One possibility is that there is yet another extra extra strong force, 'extended technicolour', with a scale factor $\Lambda_5 \gg \Lambda_4$, which connects technifermions of mass $ca. \Lambda_4$ with the otherwise massless quarks and leptons. In this case the quark and lepton masses are theoretically determined to be of order Λ_4^3/Λ_5^2, from which we infer that Λ_5 is in the multi-TeV range.

One other observable parameter must be mentioned, the Newton constant of gravitation G. An increasingly popular view (which I share) is that this is not a fundamental constant, but related to an energy scale M_G at which some entirely new physics enters. On this view, the effective Lagrangian that describes gravitation at ordinary energies is a power series

$$\mathscr{L}_{\text{eff}} = \sqrt{g}[c_0 M_G^4 + c_1 M_G^2 R + c_2 R^2 + c_2' R^{\mu\nu} R_{\mu\nu} + c_3 M_G^{-2} R^3 + \ldots],$$

with dimensionless constants c_i. If $c_1, c_2, c_2', c_3, \ldots$ are of order unity, then the only terms we would have had any chance of observing experimentally are the c_0 and c_1 terms, leaving us with conventional general relativity plus a cosmological constant. If we define M_G so that $c_1 \equiv 1$, then from the experimental value of G we conclude that $M_G = (16\pi G)^{-\frac{1}{2}} = 1.72 \times 10^{18}$ GeV. Unfortunately c_0 turns out not to be quite of order unity; from upper limits on the cosmological constant, we conclude that $c_0 < 10^{-119}$. No one knows why.

Not only does the QCD (and technicolour and extended technicolour) coupling vary with energy: the same is true of the electroweak couplings g_1 and g_2, but with a slower rate of variation. When g_1, g_2, and g_3 are extrapolated to very high energy, they are found to come together (with relative normalizations fixed by the menu of quarks and lepton quantum numbers) at an energy of order 10^{15} GeV. This allows one plausibly to suppose that the strong and electroweak gauge groups are subgroups of some 'grand unified' gauge group. On this view, it is the symmetry breaking scale M_{GUT} that determines the point where g_1, g_2, and g_3 come together, so M_{GUT} must be of order 10^{15} GeV, and it is the GUT coupling g_{GUT} that provides the common value of g_1, g_2, and g_3 at this energy. The fact that Λ_1, Λ_2, and Λ_3 are enormously different from each other and from M_{GUT} is then simply explained by the smallness of g_{GUT}. Unfortunately, in the case of electroweak symmetry breaking by elementary scalars, it is a mystery why the vacuum expectation value $\langle \phi \rangle \approx 300$ GeV is so much less than M_{GUT}. This is the well known hierarchy problem. Another problem that is not so often mentioned is why the renormalization scale Λ_{GUT} of the grand gauge group is so different from M_{GUT}, or in other words, why is g_{GUT}^2 so small?

The only theories I know that provide much hope of being able to calculate gauge couplings like g_{GUT} from fundamental principles are those that derive the gauge couplings from gravity in 6 or more dimensions. An extended version of the portion of this talk that dealt with this topic will be published in the proceedings of the Fourth Workshop on Grand Unification and the Shelter Island II Conference, so I will not summarize it here.

Gravitation does not seem to admit a simple quantum-mechanical description, whether in four or more dimensions. As already mentioned, it seems likely that really new physics enters at the Planck scale of 1.7×10^{18} GeV, and this will involve new dimensionless ratios of coupling parameters. I think it is likely that all these apparently fundamental constants will ultimately be determined by a condition of consistency of quantum mechanics with relativity, a condition that requires that all couplings have values that place the theory on a trajectory that is attracted to an ultraviolet fixed point.

Research supported in part by the Robert A. Welch Foundation and under N.S.F. contract PHY-82-15249.

Discussion

H. B. NIELSEN (*Bohr Institute, Copenhagen, Denmark*). Professor Weinberg seems to cling rather strongly to wanting a continuum theory rather than one with a fundamental cut-off (a lattice or whatever), since he even wants coupling parameters to be adjusted to an ultraviolet fixed point to make the continuum work. Would Professor Weinberg not believe a fundamental cut-off theory?

S. WEINBERG. I cannot say whether I would believe a fundamental cut-off theory until I see one. I shall just say that I hope that whatever cut-off is provided by nature will not work too well. It would be a great pity if such a cut-off would allow us to formulate theories with arbitrary coupling parameters, because then we would then have no idea of what it is that determines these parameters, apart perhaps for ideas like the anthropic principle. The great thing about the requirement of attraction to an ultraviolet fixed point is that in principle it can determine all or all but a finite number of coupling parameters.

J. G. TAYLOR (*Department of Mathematics, King's College London, Strand, London WC2R 2LS, U.K.*). After this most exciting idea of Professor Weinberg's to calculate the gauge coupling constant in higher-dimensional theories there is still the problem of ultraviolet divergences at higher loops hanging over it all, especially those arising from the gravitational radiative corrections. It is known that higher dimensional theories have far worse ultraviolet divergences than lower dimensional ones. Thus unextended supersymmetric Yang–Mills theories have infinite S-matrix elements even at 1 loop (Green & Schwarz 1982; Ragiadakos & Taylor 1983), whereas recently the dimensionally reduced theory in 4 dimensions, $N = 4$ super-Yang–Mills, is finite to all orders (Mandelstam 1982; Howe *et al.* 1982). It may then be dangerous to assume that the quantum fluctuations in higher-dimensional theories will allow neglect of higher-loop orders in some situations even if such neglect may be ultimately justified in a strictly four-dimensional theory. It may also be incorrect to assume that renormalization-group arguments, now restricted to a finite-dimensional subset of the parameter space, will be satisfactory in the higher (space–time) dimensional theories considered by Professor Weinberg.

S. WEINBERG. In the work on higher-dimensional theories discussed here, the neglect of higher loops was justified as a consequence of the presence of a large number of matter fields. This justification is of course not rigorous, and in any case may not apply in the real world. However, I do not think that the problem is any worse in six or more dimensions than in four. A non-renormalizable theory is just a field theory with an infinite number of coupling parameters, all those needed to provide counterterms to the ultraviolet divergences. The neglect of higher loops is justified in such a theory with large numbers of matter fields as long as we assume that the higher coupling parameters are not much larger than would be generated by radiative corrections. Also, the renormalization group approach applies here as it does in renormalizable theories. But at any rate, we do not yet have a four-dimensional renormalizable theory of gravitation, nor in my view are we likely ever to have one.

References

Green, M. & Schwarz, J. 1982 In *Proceedings of the International High Energy Physics Conference, Paris*. Geneva: CERN.
Howe, P., Stelle, K. S. & Townsend, P. 1982 In *Proceedings of the International High Energy Physics Conference, Paris*. Geneva: CERN.
Mandelstam, S. 1982 In *Proceedings of the International High Energy Physics Conference, Paris*. Geneva: CERN.
Ragiadakos, C. & Taylor, J. G. 1983 *Phys. Lett.* B (In the press.)

The strong, electromagnetic and weak couplings

By C. H. Llewellyn Smith

Department of Theoretical Physics, 1 Keble Road, Oxford OX1 3NP, U.K.

A brief review, aimed at non-specialists, is given of present knowledge of the parameters needed to describe strong and electroweak interactions at energies up to a few hundred GeV. Empirical evidence that the strong and electroweak couplings converge to a single 'grand unified' coupling at much higher energies is reviewed and the uncertainties in the calculation of the proton lifetime in the minimal SU(5) theory are discussed.

1. Strong interactions

There is general agreement that strong interactions are described by quantum chromodynamics, which is the only sensible theory which fits the facts (for an introduction to QCD and references see Llewellyn Smith 1982). QCD describes the interaction of coloured quarks through the exchange of coloured spin 1 'gluons', as indicated more or less directly by the data, especially the evidence for chiral symmetry. There are also gluon self interactions, which are required for the theory to make mathematical sense.

The lagrangian of QCD appears to contain several parameters: the strong coupling constant (g_s), quark masses (m_q) and a mysterious parameter θ. I shall ignore θ, which is known to be less than 1.5×10^{-9}; presumably there is some principle, outside QCD, which forces it to be very small or zero. In considering those hadrons whose flavour quantum numbers are carried by u and d quarks (i.e. with strangeness, charm, etc., equal to zero), it is likely that we can ignore the existence of heavier quarks to a good approximation. It is also an excellent approximation to set $m_{u,d} = 0$. In this case the Lagrangian exhibits exact chiral symmetry which is realized in the Nambu–Goldstone mode, leading to $m_\pi = 0$ (a measure of symmetry breaking is $m_\pi^2/m_\rho^2 = 0.03$) and many other predictions which work to better than 10%.

With just u and d and $m_{u,d} = 0$, the only remaining parameter is g_s. However, g_s depends on the energy, E, at which it is measured. Consequently we can replace the dimensionless parameter g_s by a parameter E_0 with units of energy or inverse length (in the usual $\hbar = c = 1$ units), for example, we can define E_0 by $g_s(E_0) \equiv 1$. However, as there are no other dimensional quantities in the theory, E_0 is not actually a parameter, but simply the unit of energy. Everything else is calculable in principle in terms of E_0, e.g. m_p/E_0, m_ρ/E_0 are calculable and so, therefore, is $g_s(m_p)$! Thus, with only u and d quarks QCD has *no* parameters in the chiral limit!

The idea that couplings such as g_s are energy dependent is so important for QCD and for attempts at grand unification, discussed below and in the accompanying talks by Weinberg and Ellis, that I shall indicate briefly how it comes about. Consider a classical test charge placed in a dielectric. Its charge $Q(R)$, defined in terms of the electric flux out of a sphere surrounding it, depends on the radius, R, of the sphere. For R much smaller than the typical intermolecular distance it will have the same value, Q, as in free space, but at larger distances there is screening, due to a net flow of negative polarization charge into the sphere, and $Q(R)$ decreases to Q/ϵ. Likewise in field theory vacuum polarization makes charges distance (or equivalently energy) dependent. In QED, for example, the contribution of virtual particles (through e⁺e⁻ loops, etc.)

to the photon propagator causes the effective value of α, which controls the electromagnetic force between two charged particles, to increase from its value at infinity of $(137.036...)^{-1}$ as the particles approach one another. For small four-momentum transfer, q^2, the dominant terms give

$$\alpha_{\text{eff}}(q^2) = \alpha - (\alpha^2/15\pi)(q^2/m_e^2),$$

which contributes a very well verified -27 MHz to the Lamb shift.

A similar phenomenon occurs in all field theories, except that in the case of non-Abelian theories such as QCD the coupling *increases* with distance or, equivalently, decreases with increasing energy (Gross & Wilczek 1973; Politzer 1973). For example, in QCD with four flavours of quarks

$$g^2(E) = 48\pi^2/(25\ln(E^2/\Lambda^2)),$$

to leading order at large E. Here we see explicitly that the value of the dimensionless quantity g is determined by a dimensional parameter Λ, which sets the scale at which strong interactions are strong (a precise definition of Λ requires that we go beyond leading order and depends on technical details of how the theory is formulated; below we quote values of $\Lambda = \Lambda_{\overline{\text{MS}}}$ for the so-called barred minimal subtraction ($\overline{\text{MS}}$) scheme).

The idea of calculating g_s, or equivalently calculating Λ/M where M is a directly observable quantity with dimensions of mass which can be used as a scale, has actually been realized in lattice QCD. In this formulation, fields are only defined at discrete points on a lattice in space–time or on the links between them (for reviews and references see Creutz *et al.* 1983; Kogut 1982; Rebbi 1982), the intention being to let the lattice spacing, a, tend to zero at the end of the calculation. For finite a, the theory can be simulated on a computer. So far, only the gluons have been treated as dynamical degrees of freedom in most calculations. Heavy quarks can be introduced as static sources and the calculations indicate that the force between them corresponds to a quark-confining potential $V = \sigma R$ at large R even for $a \to 0$. For given a, the coupling $g(a)$ is adjusted to give $\sigma \approx (0.43\,\text{GeV})^2$, as required by the spectroscopy of heavy quark systems, or $\sigma \approx (0.50\,\text{GeV})^2$, which gives the observed slope of Regge trajectories in the string model. The quantity that corresponds to $g_s(E)$ in the continuum can be calculated in terms of $g(a)$ and it is found (Creutz *et al.* 1983) that

$$\Lambda_{\overline{\text{MS}}} = (0.23 \pm 0.05)\,\sigma^{\frac{1}{2}} \approx 110\,\text{MeV},$$

thus giving $g_s(E)$ *absolutely* ($\sigma^{\frac{1}{2}}$ serving as the unit of energy). There has also been some encouraging progress in calculating hadron masses on the lattice (see Creutz *et al.* 1983; Rebbi 1983) but the overall size of the lattices used so far has been too small (typically 1 fm[†] or less compared with the root mean square radius of the proton which is 0.8 fm) for the results to be expected to be realistic. Furthermore the calculations do not include 'dynamical' fermions in virtual loops (although there are some arguments which suggest that their effects will be small).

The value of $\alpha_s(E)$ can be measured in some high energy experiments for which predictions can be made with QCD perturbation theory (for a recent review with references to the original papers see Altarelli 1982). Large E is a necessary condition for perturbation theory to be used but in general it is not sufficient. Mathematically, the perturbation expansion is spoiled by terms such as $(\alpha_s(E)\ln(E/m))^n$ in most cases; the divergence as $m \to 0$ shows physically that the result is sensitive to long distance physics so that it depends on how hadrons are constructed from quarks and gluons, which is obviously not described by perturbation theory. However, there are some processes for which factors of $\ln E/m$ do not occur that are insensitive to how hadrons are made of

[†] 1 fm = 10^{-15} m.

quarks and gluons (e.g. σ ($\bar{e}e \to$ hadrons) and the Q^2 dependence of deep inelastic structure functions) which can presumably be treated perturbatively. It is hard to extract α_s from the data for these processes as it is not very small, so perturbation theory converges slowly, and it is even harder to extract Λ because α_s is only sensitive to Λ at small E, where it is large and there are important incalculable subasymptotic contributions, of relative size μ^2/E^2. Various experiments, whose results all agree qualitatively at least with the predictions of perturbative QCD, give $\Lambda_{\overline{\text{MS}}}$ in the range 100–350 MeV (for recent results see Eisle 1982), not in disagreement with the lattice calculations.

To conclude on the strong interactions: QCD is also certainly correct. In the light quark sector, it has no parameters in the limit $m_{u,d} = 0$. The quark masses (for light and heavy quarks) must be introduced as parameters from outside QCD: the big question being, what determines m_q/Λ? The main challenge for theorists is to show that QCD really leads to the known hadrons with the properties observed.

2. Electroweak interactions

We can no longer discuss the electromagnetic and weak interactions separately. Indeed, the differential cross section for $e^+e^- \to \mu^+\mu^-$, long thought of as a testing ground for QED, disagrees with pure QED at high energies, although the departure from pure QED is well described by interference with the expected neutral current contribution (for a review see Davier 1982).

The parameters in the standard $SU(2) \times U(1)$ electroweak gauge theory fall into two categories (for a recent review of electroweak gauge theories see Beg & Sirlin 1982):

(i) the gauge couplings g_1 and g_2 or, equivalently, α_{em} and $\sin^2 \theta_W$;

(ii) parameters associated with symmetry breaking, i.e. couplings of Higgs bosons in the canonical model: the Higgs self couplings, which determine m_H and $\langle \phi \rangle$ (which in turn determines m_W and m_Z), and the Yukawa couplings which determine the quark masses and Cabbibo–Kobayashi–Maskawa mixing angles. I shall assume the simplest possible symmetry breaking scheme in which $m_W = m_Z \cos \theta_W$ to lowest order.

To lowest order, processes which do not involve the Higgs boson directly can be described by three parameters, in addition to fermion masses and mixings, which can be chosen to be α, $\sin^2 \theta_W$ and the Fermi constant G_F. A very large amount of data is well described in terms of these parameters. In particular, different experiments give consistent values of $\sin^2 \theta_W$; the values determined by the most accurate experiments are shown in the accompanying table. As stressed particularly by Veltman (Veltman 1980; Green & Veltman 1980) a vital question is whether this agreement survives electroweak radiative corrections; tests which are sensitive to second and higher order effects are the electroweak equivalent of the Lamb shift, corrections to the muon's g factor and the other classic tests of QED (for reviews see Wheater 1982; Aoki et al. 1982). Nominally, second order effects shift $\sin^2 \theta_W$ by an amount of order ± 0.02; in fact detailed calculations give shifts which are somewhat less for the experiments shown in the table and do not destroy the agreement between different measurements (beyond the lowest order the value of the parameter $\sin^2 \theta$ depends on how it is defined; the values in the table are for the $\overline{\text{MS}}$ scheme with scale m_W). Another way to express the magnitude of second order effects is to compare the values $m_W = 78.2^{+2.7}_{-2.5}$ and $m_Z = 89.0^{+2.2}_{-2.0}$ derived from $\nu N \to \nu X$ in lowest order and $m_W = 83.1^{+3.1}_{-2.8}$, $m_Z = 93.8^{+2.5}_{-2.2}$ derived taking electroweak corrections to the experiment and to the mass formula into account (Wheater & Llewellyn Smith 1982). Clearly the experiments are becoming sensitive to second order effects† and incisive tests of $SU(2) \times U(1)$ will be possible in the next few years with $\nu e \to \nu e$, $\nu N \to \nu X$, $ep \to eX$, $e^+e^- \to \mu^+\mu^-$, m_W, m_Z among other quantities.

These measurements probe the idea that the W and Z are fundamental gauge bosons. If they are composite the same predictions can be obtained to leading order in a large class of theories (Bjorken 1979) but the properties of the W and Z would differ to second order, unless the binding

TABLE 1. VALUES OF $\sin^2 \theta_W$

(Determined (a) with the Born approximation for the electroweak interactions (column 1) and (b) by including one loop corrections in the $\overline{\text{MS}}$ scheme with scale m_W (column 2); the corrected values are taken from Wheater & Llewellyn Smith (1982) for νN and ed (νN has also been treated by Marciano & Sirlin (1981) with the same result). The corrections for m_W are based on Marciano & Sirlin (1980). The Born value for νN is based on an average of experiments. The value for ed is from the fit of Kim et al. (1981) to the experiment of Prescott et al. (1979). The value of m_W is from Arnison (1983a).)

	$\sin^2 \theta_W$	$\sin^2 \overline{\theta}_W(m_W)$
$\nu N \to \nu X$	0.227 ± 0.015	0.215 ± 0.015
$e_L^- d - e_R^- d$	0.223 ± 0.015	0.215 ± 0.015
$m_W = 81 \pm 5$	0.22 ± 0.03	$0.226^{+0.030}_{-0.026}$

energy is very large. These measurements are also sensitive to the contributions of new relatively light particles, such as the plethora of new particles with masses of order m_W whose existence is predicted in supersymmetric theories, which would alter the radiative corrections at the per cent level (for details see Schwarzer 1983).

To conclude on $SU(2) \times U(1)$: it is certainly correct to first approximation and even more incisive tests will soon be possible. However the conventional symmetry breaking mechanism (the Higgs effect) seems very *ad hoc* and is prolific in arbitrary parameters (m_W, m_Z, m_H, m_{q_i}, θ_{KM}). The challenging question is what fixes these parameters and why are the masses so disparate ($m_W/m_e \approx 1.6 \times 10^5$; $m_e/m_t < 2.5 \times 10^{-3}$, etc.)?

3. GRAND UNIFICATION?

The idea that $SU(3)_c$ (the gauge group of QCD), $SU(2)$ and $U(1)$ are subgroups of a single 'grand unifying' group is very appealing (for a review and references see Langacker 1981). If indeed strong and electroweak interactions are fundamentally the same, the definition of a hadron or lepton has no fundamental significance and it is natural to attempt to unify quarks and leptons by putting them in the same representation of the gauge group. If we assume that the fifteen states that form the 'first generation' of fermions (e_L^-, e_L^+, ν_L, $u_L^{r,b,g}$, $u_L^{r,b,g}$, $d_L^{r,b,g}$, $\bar{d}_L^{r,b,g}$) form one, possibly reducible, representation it would follow that

$$g_2 = g_3,$$

if the symmetry were unbroken, where $g_{2,3}$ are the $SU(2)$ and $SU(3)$ couplings, and

$$g_1 \equiv (\tfrac{5}{3})^{\frac{1}{2}} g' = g_2 \quad \text{or} \quad \sin^2 \theta_W = \tfrac{3}{8},$$

where g' is the conventionally defined $U(1)$ coupling (Georgi et al. 1974). The 'grand symmetry' G must be broken by the gauge bosons corresponding to those generators of G which do not belong to $SU(3) \times SU(2) \times U(1)$ acquiring very large masses (which we shall call generically m_X).

† Since this talk was given, the discovery of the Z was announced with (based on 5 events) $m_Z = 95.2 \pm 2.5$ GeV and an improved value of $m_W = 81 \pm 2$ GeV was given, based on 27 events (Arnison 1983b); the experimenters stress that final calibration of the calorimeter is still in progress and small scale shifts in these masses, most likely affecting both, are still possible.

These vector bosons couple quarks to leptons and mediate nucleon decay, to which we return later. For energies which are asymptotically large compared with m_X, the symmetry becomes exact. Thus the couplings behave as shown in figure 1. Note that g_1/g_2 decreases with energy so that
$$\sin^2\theta(E) = 3g_1^2(E)/(3g_1^2(E) + 5g_2^2(E))$$
will be less than the symmetry value of $\tfrac{3}{8}$ at low energy.

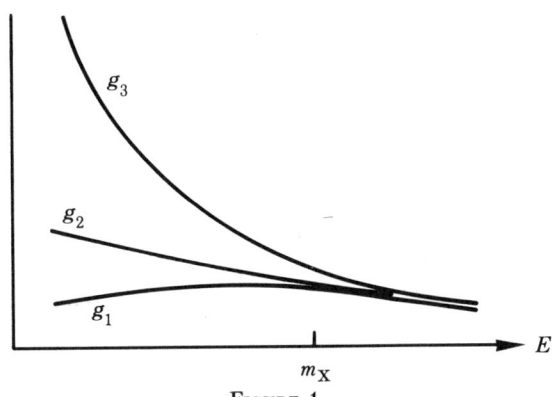

FIGURE 1

The precise way that the g_i approach each other for $E \gtrsim m_X$ depends on G. However, as a first approximation we can set $m_X = 0$ for $E > m_X$ and $m_X = \infty$ for $E < m_X$. This approximation, in which the g_i meet at $E = m_X$ and their evolution up to this point can be studied without knowledge of G (Georgi et al. 1974), gives $\sin^2\theta \approx 0.21$ at low energy, in good agreement with experiment, and m_X of order 10^{15} GeV.

To find more than the order of magnitude of m_X, and to obtain $\sin^2\theta_W$ precisely, it is necessary to specify the group G. The simplest choice is SU(5) (Georgi & Glashow 1974). The minimal version of this theory, with no ingredients for which there is no phenomenological necessity, gives
$$\sin^2\theta_{\overline{\mathrm{MS}}}(m_W) = 0.215 \pm 0.006,$$
for $\Lambda_{\overline{\mathrm{MS}}} = 150^{+250}_{-100}$ MeV, in excellent agreement with the experimental value given above, and $m_X = (1.3^{+0.9}_{-0.6}) \times 10^{15}\Lambda_{\overline{\mathrm{MS}}}$ (these quantities have been calculated independently by many authors; here we quote the results of Llewellyn Smith et al. 1981). Uncertainties in low energy data, used in evaluating the evolution of α_{em}, contribute ± 0.18 to the error in the coefficient 1.3; the error from 'three loop' contributions, which have not been calculated, is taken to be ± 0.15, which is probably reasonable as they are proportional to $\alpha_s(m_W)/\pi$; a further error of $\pm 30\%$ is introduced by allowing the very massive coloured Higgs bosons to range from $10^{-2}m_X$ to $10^{+2}m_X$. Given a value for m_X, we can calculate the nucleon decay rate using a model to calculate the matrix elements of the appropriate four fermion operator. We consider the decay $\mathrm{p} \to \pi^0 \mathrm{e}^+$, for which the best experimental limits exist, which is thought to be the dominant mode in SU(5). Of all credible model calculations, the quark model gives the longest lifetime but even so it is possible that it yields an underestimate. Somewhat arbitrarily introducing an error to cover an underestimate by a factor of two gives:

$$\Gamma(\mathrm{p} \to \pi^0 \mathrm{e}^+)^{-1} = (3^{+3}_{-?}) \left(\frac{m_X}{1.3 \times 0.15 \times 10^{15}\,\mathrm{GeV}}\right)^4 \left(\frac{\ln(m_N/\Lambda)}{\ln(m_N/0.15)}\right)^{\tfrac{12}{21}} \times 10^{29}\,\text{years}$$

$$= (6^{+90}_{-?}) \left(\frac{\Lambda_{\overline{\mathrm{MS}}}}{350\,\mathrm{MeV}}\right)^4 \times 10^{30}\,\text{years},$$

where the first line is normalized to $\Lambda_{\overline{\text{MS}}} = 150$ MeV, for some time the standard theoretical guess, and the second to 350 MeV because there is some evidence that $\Lambda_{\overline{\text{MS}}} > 150$ MeV and 350 GeV is about the largest value currently thought to be acceptable (the median value is based on the quark model calculation of Isgur & Wise (1982)). We see that if $\Lambda_{\overline{\text{MS}}} \approx 350$ MeV, the current results (Goldhaber 1983) do not quite rule out SU(5). However, a limit greater than 10^{32} years could only be accommodated by a conspiracy of the uncertainties, and an even larger value of $\Lambda_{\overline{\text{MS}}}$, and would essentially eliminate the minimal version of SU(5).

Longer lifetimes can easily be obtained by changing the model, for example to SO(10), or to a supersymmetric GUT. However, if this is done the precise prediction for $\sin^2\theta$ is lost; the predicted value is still of order 0.21 but it could be bigger or smaller by as much as ± 0.05, depending on new unknown parameters, although the measured value can still be accommodated in most models.

To conclude on grand unification: the underlying idea is very appealing and it is supported by the fact that the couplings do seem to merge at energies of order 10^{15} GeV. The minimal SU(5) model is strikingly successful in explaining the value of $\sin^2\theta_W$ but it predicts a rate for $p \to \pi^0 e^+$ which seems to be on the verge of being ruled out. The lifetime can be made longer by altering the theory but this opens a Pandora's box of more complex alternatives.

4. General conclusions

The evidence for $SU(3) \times SU(2) \times U(1)$ is excellent: QCD is correct and the electroweak interactions are unified. However, this successful model is surely not the final theory: it has far too many arbitrary parameters, it does not explain the different 'generations' and it does not incorporate gravity. In any case the big problem for any more complete theory is the origin of masses, or equivalently the origin of symmetry breaking, and the origin of enormous ratios of masses such as m_W/M_P, where M_P is the Planck mass of 10^{19} GeV which characterizes gravity, or m_W/m_X.

References

Altarelli, G. 1982 *Physics Rep.* **81**, 1.
Aoki, K., Hioki, Z., Kawabe, R., Konuma, M. & Muta, T. 1982 *Prog. theor. Phys.* (Supplement 73.)
Arnison, G. et al. 1983a *Physics Lett.* B **122**, 103.
Arnison, G. et al. 1983b *Physics Lett.* B **126**, 398.
Beg, M. A. & Sirlin, A. 1982 *Physics Rep.* **88**, 1.
Bjorken, J. D. 1979 *Phys. Rev.* D **19**, 335.
Creutz, M., Jacobs, L. & Rebbi, C. 1983 *Physics Rep.* **95**, 201.
Davier, M. 1982 *J. Phys., Paris* C **3**, 723.
Eisle, F. 1982 *J. Phys., Paris* C **3**, 337.
Georgi, H. & Glashow, S. L. 1974 *Phys. Rev. Lett.* **32**, 438.
Georgi, H., Quinn, H. R. & Weinberg, S. 1974 *Phys. Rev. Lett.* **33**, 351.
Goldhaber, M. 1983 *Phil. Trans. R. Soc. Lond.* A **310**, 225.
Green, M. & Veltman, M. 1980 *Nucl. Phys.* B **169**, 137 (E: B **175**, 547).
Gross, D. & Wilczek, F. 1973 *Phys. Rev. Lett.* **30**, 1343.
Isgur, N. & Wise, M. B. 1982 *Physics Lett.* B **117**, 179.
Kim, J. E., Langacker, P., Levine, M. & Williams, H. H. 1981 *Rev. mod. Phys.* **53**, 211.
Kogut, J. B. 1983 *Rev. mod. Phys.* **55**, 775.
Langacker, P. 1981 *Physics Rep.* **72**, 185.
Llewellyn Smith, C. H. 1982 *Phil. Trans. R. Soc. Lond.* A **304**, 5.
Llewellyn Smith, C. H., Ross, G. G. & Wheater, J. F. 1981 *Nucl. Phys.* B **177**, 263.
Marciano, W. J. & Sirlin, A. 1980 *Phys. Rev.* D **22**, 2695.
Marciano, W. J. & Sirlin, A. 1981 *Nucl. Phys.* B **159**, 442.

Politzer, H. D. 1973 *Phys. Rev. Lett.* **30**, 1346.
Prescott, C. Y. *et al.* 1979 *Physics Lett.* B **84**, 524.
Rebbi, C. 1983 *J. Phys., Paris* C **3**, 723.
Schwarzer, K. 1983 Oxford preprint. (In preparation.)
Veltman, M. 1980 *Physics Lett.* B **91**, 95.
Wheater, J. F. 1982 *J. Phys.* C **3**, 305.
Wheater, J. F. & Llewellyn Smith, C. H. 1982 *Nucl. Phys.* B **208**, 27.

Discussion

H. B. NIELSEN (*Bohr Institute, Copenhagen, Denmark*). How strong evidence for unification is it that the couplings of the gauge theories agree so well? Could one not obtain a similar prediction by having all three coupling constants coming for instance from a scheme like the one Steven Weinberg told us about? The prediction that the couplings are equal even might be dreamt to be not too difficult if one has any way of predicting them at all.

C. H. LLEWELLYN SMITH. The fact that, suitably normalized, the couplings seem to converge at high energy provides evidence for a connection between the different interactions. It is of course possible that this connection has nothing to do with conventional grand unified theories.

Field theories without fundamental gauge symmetries

By H. B. Nielsen

*The Niels Bohr Institute, University of Copenhagen and NORDITA,
Blegdamsvej 17, DK-2100 Copenhagen Ø, Denmark*

By using the lack of dependence of the form of the kinetic energy for a nonrelativistic free particle as an example, it is argued that a physical law with a less extended range of application (non-relativistic energy momentum relation) often follows from a more extended one (in this case the relativistic relation) without too much dependence on the details of the latter. We extend the lesson from such examples to the ideal of random dynamics: no fundamental laws are needed to be known. Almost any random fundamental model will give the correct main features for the range of physical conditions accessible to us today (energies less than 1000 GeV) even if it is wrong in detail.

This suggests the programme of attempting to 'derive' the various symmetries and other features of physics known today from random models at least without the feature to be derived.

As an example, D. Förster, M. Ninomiya and myself 'derive' gauge invariance in this way (Förster *et al.*, *Phys. Lett.* B **94**, 135 (1980)), and show that it has at least a non-zero probability for being effectively a symmetry. In fact we show that a certain non-gauge-symmetric lattice model has zero mass photons for a whole range of its parameters, so that it is not necessary to fine tune it to get massless photons. It comes about by means of a formal gauge symmetry achieved by introducing a superfluous number of field variables.

The achievements in our programme of random dynamics up till now are briefly reviewed. In particular, Lorentz invariance may be understood as a low energy phenomenon (S. Chadha, M. Ninomiya and myself).

An analogy between the development of physics as one goes to lower and lower energies and that of living species through the history of the Earth is put forward.

1. Introduction

The work that I shall present is part of an ambitious programme of random dynamics in which, in collaboration with various colleagues, I attempt to derive the laws of nature as we know them from fundamental laws that are assumed to be random and thus extremely complicated. (For a recent review see Nielsen 1983). I shall give as an example a piece of research work by D. Förster, M. Ninomiya and myself (Förster *et al.* 1980) in which we in a sense derive a symmetry law: gauge invariance (the same work also has been done by Shenker (1980) in an unpublished research report).

The picture of physics which I have in mind in this project of random dynamics is the following.

There exists some system of fundamental physical equations (or a fundamental action or the like) governing the time development of some fundamental fields. It may be difficult to say exactly in what terms it should be formulated, and it is part of my point that it may not be important to know this. Contrary to the speculation – that many physicists believe – that fundamental physics is simple, these fundamental equations are assumed here to be extremely complicated. Because of the high degree of complication assumed of the fundamental equations (the fundamental laws of nature we could say) we have to give up any hope of guessing their exact form. Our best hope is, then, to guess a very large class of possible fundamental equations (or actions) and a probability measure over that class. It would then make sense to assume that the actual fundamental equation system (or action or whatever) is randomly chosen from that class in the sense that after having added assumptions about how to connect the fundamental fields (degrees of freedom) to experimental observations one would find agreement with experiment within statistically expected accuracy.

At the present stage of science we have a series of regularities and theories that are well tested in some regions of application (energies per particle less than those corresponding to highest accelerator energies).

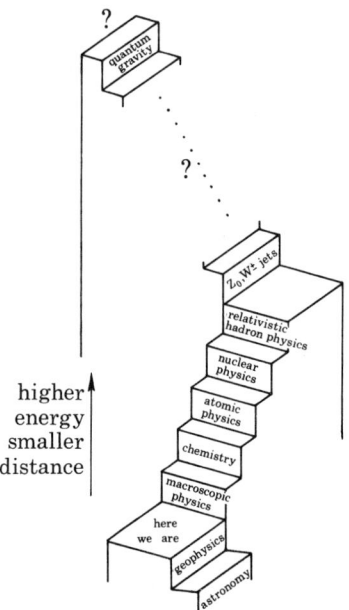

FIGURE 1. The quantum staircase illustrating how the various branches of physics – corresponding to the steps in the staircase – may be roughly considered as characterized by various scales of energy or inverse length (we put Planck's constant h and c equal to unity).

These regularities, such as the principle of relativity, may be called laws of nature, but we do not think of them as fundamental in the picture of random dynamics. Rather they should be derived in some limit as consequences of the fundamental laws of nature (i.e. the fundamental equation system or action or whatever) which are supposed random. This means that the success of the random dynamics idea hinges upon the possibility of deriving the 'non-fundamental' laws of nature from a major part, that is a part with high probability, of the complicated fundamental equation systems in the assumed large class.

In practice we concentrate on explaining the (presumed not so fundamental) regularities (or laws of nature) already known. But now instead of attempting to derive them from a totally random system of equations or actions we would in practice take a system (a model) in which several regularities other than the one to be derived are already present. If it is possible to derive a regularity from a random action without any regularities it is normally easier to do so if one is allowed to use some principles that are already known.

As we progress we may hope to use successively fewer and fewer known principles, and hope,

[52]

FIELD THEORIES WITHOUT GAUGE SYMMETRIES

at the end of a long series of calculations, to abandon any assumption about the existence of time or that the dynamics should be given by a Lagrangian. We may hope to go that far some day but it is difficult to imagine even in which language to formulate a model without time and without an action principle.

It may not be good to use the phrase 'laws of nature' since it means two things: (1) a regularity, a principle such as rotational invariance, the principle of relativity, or the linearity between energy and momentum squared for a non-relativistic free particle; (2) a complete model of describing in detail the dynamics for a branch or the whole of physical phenomena.

In both of these definitions the physical law can be either fundamental or only (approximately) valid under some limited range of physical conditions (for example when no particle has more energy than 1 TeV).

According to the hypothesis of random dynamics there may exist a fundamental law of nature in sense (2) of a model although it is so complicated that we shall not attempt to guess it. In this sense of a physical law (a dynamical model) there is no contradiction in taking the law to be random. With the other meaning (1), that of a regularity, a random law would hardly make sense, but according to 'random dynamics' there should not exist any fundamental law of nature in this sense.

Many non-fundamental laws of nature in sense (1) (of regularities) are known. In sense (2) (of a model), the standard model, i.e. QCD combined with the Weinberg–Salam–Glashow electro-weak gauge model, is one that presumably explains all we know today except for gravitational phenomena.

In the following section I shall put forward the possibility that the fundamental laws of nature are random and so complicated that we could not attempt to guess them, using as an example a derivation of the kinetic energy of a non-relativistic particle.

In §3 we go to a lattice field theory model without any *a priori* gauge invariance. Nevertheless, it turns out that one can change the notation so that a formal gauge symmetry arises. In spite of this symmetry being formal it turns out that there is some chance that it can provide an explanation for the light quantum (the photon) being massless, a phenomenon due to gauge invariance.

Finally, in §4 I present a very condensed list of what we have achieved or are about to achieve in the random dynamics programme and give concluding remarks in §5.

2. THE PROGRAMME OF RANDOM FUNDAMENTAL LAWS OF NATURE

In the random dynamics programme we wish to derive the regularities known today, such as Lorentz invariance and gauge symmetry, but which we suppose not to be fundamental, from a more fundamental model. We must admit that we do not know the more fundamental model but rather we shall choose a large class of such models and a probability measure on the latter; then we assume that a random model is valid.

Now I would like to illustrate how a regularity – the linearity between momentum squared p^2 and energy $E(p)$ for a free non-relativistic particle – can arise in a limit, that of low velocity, in a case where we also know the more broadly valid model, which in this case is special relativity. We consider this example of random dynamics, which could have been done before the advent of special relativity, both for illustration and to argue a case where we know, i.e. after Einstein, that the idea of a derivation in accordance with the random dynamics programme is indeed correct. The relation between energy E and 3-momentum $\boldsymbol{p}$ for a particle according to the special theory of relativity is shown in figure 2,

$$E = (\boldsymbol{p}^2 + m^2)^{\frac{1}{2}} \tag{2.1}$$

in units wherein the velocity of light, c, equals one. A research project of the random dynamics type which could be imagined before Einstein would be to 'derive' the non-relativistic energy momentum relation

$$E = \boldsymbol{p}^2/2m \quad \text{or} \quad E \propto \boldsymbol{p}^2. \tag{2.2}$$

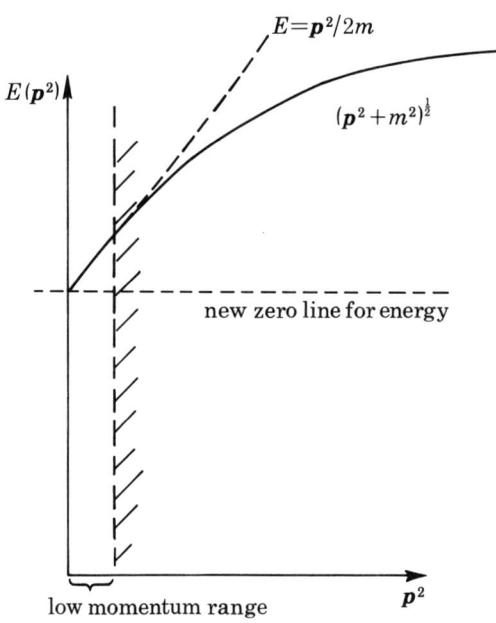

FIGURE 2. The relation between energy and (3-)momentum squared for a free particle. In non-relativistic physics (nuclear physics say) we only have access to the small momentum region where we can approximate the curve by a portion of a straight line.

Here m is a constant, the mass. This could be done from a 'random model' in which

$$E = f(\boldsymbol{p}^2), \tag{2.3}$$

and the function f is the object that is chosen randomly out of a large class of functions, say the analytic ones.

Even though in the derivation of $E(\boldsymbol{p}) \propto \boldsymbol{p}^2$ we use a random model that turns out later to be the special theory of relativity, it does not mean that, if I had done this last century, I should have necessarily believed that this random model was the most fundamental one. It might well be that what is used at one level as fundamental is just an approximation at another level.

We could argue from the principle of translational invariance that the energy E of a free particle cannot depend on the position but only on the momentum $\boldsymbol{p}$. We could further argue from rotational invariance for it being only a function of $\boldsymbol{p}^2$ and thus of the form (2.3). It is typical in practice in research work in the random dynamics programme to assume some already known principles although one must of course avoid assuming what we want to 'derive'.

Almost any analytic function $f(\boldsymbol{p}^2)$ would have a Taylor expansion

$$f(\boldsymbol{p}^2) = f(0) + f'(0)\boldsymbol{p}^2, \tag{2.4}$$

valid for small p, and apart from the unimportant constant $f(0)$ (if the particle is not annihilated) it has the form (2.2) if we identify
$$f'(0) = 1/2m. \tag{2.5}$$
When I say that 'almost' all analytic functions provide the same form for the energy for low momentum as the non-relativistic expression (2.2), I have in mind the possibility that the Taylor expansion coefficient $f'(0)$ could be zero. But if one imagined that the energy momentum law (2.3) was described by a randomly chosen function f, it would with a reasonable probability distribution, be very unlikely for $f'(0)$ to be exactly zero. If one has a real valued random variable, such as $f'(0)$ would be if the function f were random, and it has a smooth probability distribution, the probability for it taking any special value, e.g. $f'(0) = 0$, is zero; i.e. it is unlikely for it to take any given value specified in advance. We may say that it is unlikely that $f'(0)$ should have the fine-tuned value zero.

We thus see that the specific form (2.1) is not important for 'deriving' the low momentum or low velocity form. In fact an essentially random functional form for f would suffice.

We have here seen an example of a situation that often occurs. Details at the more fundamental level are not so important for effective physics in a corner (of low energy, say).

Thus it may be more important to assume in which corner some experiments are done than what are the fundamental equations. This makes the separation of experiment and theory less easy. See Eddington (1946); Slater (1957).

3. A LATTICE FIELD THEORY MODEL WITHOUT GAUGE INVARIANCE

As a 'modern' example of a random dynamics derivation we shall consider how Förster *et al.* (1980), and independently Shenker (1980) effectively obtained gauge symmetry.

We consider electrodynamics (Maxwell's equations) formulated in terms of the four potential $A^\mu(x) = (\Phi(x), \mathbf{A}(x))$ so that the second rank antisymmetric tensor $F_{\mu\nu}(x)$ composed from the electric field $\mathbf{E}(x)$ and the magnetic induction $\mathbf{B}(x)$ is written
$$F_{\mu\nu}(x) = \partial_\mu A_\nu(x) - \partial_\nu A_\mu(x), \tag{3.1}$$
where
$$\partial_\mu = \partial/\partial x^\mu. \tag{3.2}$$
See, for example, Bjorken & Drell (1964) for notation. One set of Maxwell's equations,
$$\partial_\mu F^{\mu\nu}(x) = j^\nu(x), \tag{3.3}$$
is then given by extremizing the action
$$S = \int -\tfrac{1}{4} F_{\mu\nu}(x) F^{\mu\nu}(x) \, \mathrm{d}^4 x + \int A_\mu(x) j^\mu(x) \, \mathrm{d}^4 x, \tag{3.4}$$
considered as a functional of A_μ, while the remaining set of Maxwell equations
$$\partial_\mu F_{\nu\rho}(x) + \partial_\nu F_{\rho\mu}(x) + \partial_\rho F_{\mu\nu}(x) = 0, \tag{3.5}$$
follows alone from the form of (3.1). The gauge symmetry is the invariance of the action (3.4) under the replacement
$$A_\mu(x) \to A_\mu(x) + \partial_\mu \lambda(x), \tag{3.6}$$
where $\lambda(x)$ is an arbitrary smooth real function defined on space–time. (Of course the four current $j^\mu(x)$ should be conserved, $\partial_\mu j^\mu(x) = 0$.)

The existence of this symmetry may be taken as the reason for the mass of the photon being zero. In fact adding a term

$$\int \tfrac{1}{2} m_\gamma^2 A_\mu(x) A^\mu(x) \, \mathrm{d}^4 x, \tag{3.7}$$

to the action S would spoil the invariance under the gauge substitution (3.6), but this term is just what would give the photon a mass m_γ.

Intuitively one thus expects that in order to obtain a model showing a zero mass photon it is necessary to fine-tune in the sense of choosing a very special combination of values for the parameters of the model. For example one would expect that the parameter m_γ should be zero, a very special value. One would think that such fine-tuning is characteristic for any model that has some symmetry.

But surprisingly enough we have found that there are several models in which symmetries do occur without fine-tuning. So the intuitive argument that a symmetry cannot appear without the fine-tuning of parameters seems to be circumvented in some cases.

In fact we (Förster et al. 1980) have considered a model that could be called 'U(1)-lattice electrodynamics without *a priori* gauge symmetry' and found that in spite of this model not being invariant under a gauge symmetry when looked at superficially we can introduce one formally and even obtain from it a physical effect, the masslessness of the photon. This model can be crudely described as the result of putting on a lattice, in a naïve way, electrodynamics with a photon mass term corresponding to the action being (3.4) plus (3.7) (the second term in (3.4) being considered a source term and therefore neglected). 'Putting on a lattice' means constructing a model in which the space(–time) continuum is replaced by a lattice of points, so that one includes, say, only those points (events) that have integer coordinates, in terms of a unit of length a called the lattice constant. Then the fields in the lattice field theory are only defined on the included points. Often though – as in the model we consider here – it is more elegant to let the field be defined on the links. The links are (see figure 3) the pieces of line connecting two neighbouring points, with integer coordinates, i.e. two points having three out of their four integer coordinates equal while one coordinate deviates by one lattice constant unit (e.g. x^ρ and $x^\rho + \delta_\mu^\rho a$). With the approximation that the lattice constant is small, the link has a rather well defined position x^ρ in space–time and it further has a direction along one of the four axes, let us say the x^μ-axis. We can then introduce in the 'naïve continuum limit' a connection between the $A_\mu(x^\rho)$-potential and the variables $U(\bullet\!\!-\!\!\bullet)$ of the model that are defined on the links and take complex values restricted to being of unit norm

$$U(\bullet\!\!-\!\!\bullet) \in \mathrm{U}(1) = \{z \in \mathbb{C} \ |z| = 1\}. \tag{3.8}$$

This connection is the relation

$$U(\overset{x^\rho}{\bullet}\!\!-\!\!\overset{x^\rho+\delta_\mu^\rho a}{\bullet}) = \exp\,(\mathrm{i}e A_\mu(x^\rho)a) \quad (e \text{ is the electric charge quantum}). \tag{3.9}$$

A technical detail is that we keep the time coordinate purely imaginary before going to the lattice.

The 'U(1)-lattice electrodynamics without *a priori* gauge symmetry' is defined by its action

$$S = \beta \sum_\square \mathrm{Re}\, U_\square + \kappa \sum_\square \mathrm{Re}\, U(\bullet\!\!-\!\!\bullet) \tag{3.10}$$

where the summation $\sum_{\bullet\!-\!\bullet}$ runs over all the links while $\sum_\square$ runs over all the plaquettes, the unit squares formed from four neighbouring links. The plaquette variable $U_\square$ is defined for each

FIELD THEORIES WITHOUT GAUGE SYMMETRIES

plaquette as the product of the four variables $U(\bullet\!\!-\!\!\bullet)$ associated with the four links making up the plaquette in question. Here one should use the convention of associating the $U(\bullet\!\!-\!\!\bullet)$ variables to oriented links in such a way that switching the orientation of the link corresponds to taking the inverse of the variable

$$U(\overset{x^\rho}{\bullet}\!\!-\!\!\overset{y^\rho}{\bullet}) = [U(\overset{y^\rho}{\bullet}\!\!-\!\!\overset{x^\rho}{\bullet})^{-1}]^{-1}, \qquad (3.11)$$

then

$$U_\square = U(\;\square\;)\, U(\;\square\;)\, U(\;\square\;)\, U(\;\square\;) \qquad (3.12)$$

is the product corresponding to the succession of oriented links around the edge of the plaquette $\square$. The dotted lines in (3.12) just symbolize the other three sides of $\square$ than the one in question.

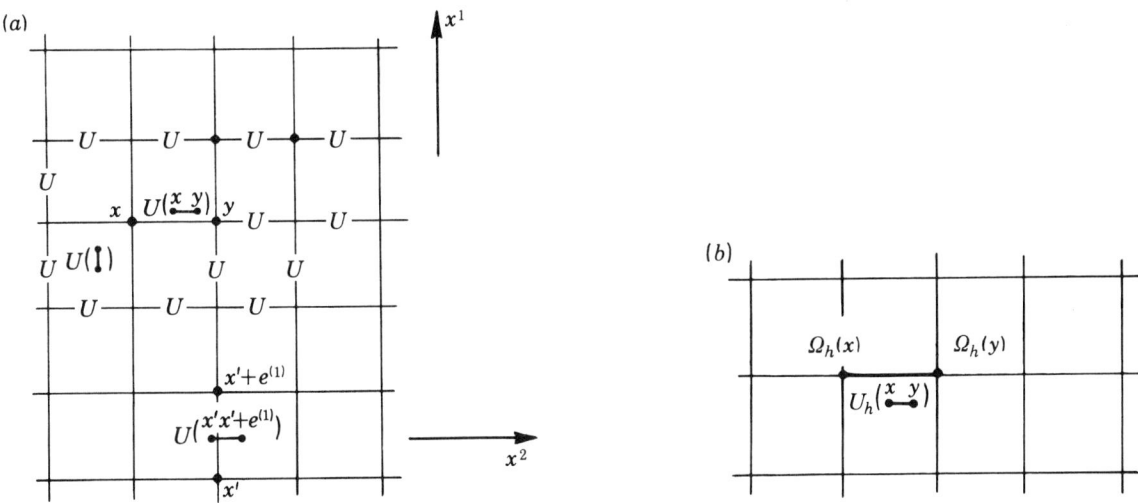

FIGURE 3. Symbolic drawings of a lattice model. In (a) we illustrate the fundamental variables $U(\bullet\!\!-\!\!\bullet)$, in (b) the variables in the formulation with superfluously many variables $U_h(\bullet\!\!-\!\!\bullet)$ and $\Omega_h(\cdot)$. The lattice should actually be 4-dimensional but has for simplicity been drawn as being 2-dimensional.

Expanding the action in a Taylor expansion to second order in $eA_\mu(x^\rho)\, a$ after insertion of (3.9), neglecting constant terms in the action and using the approximation of the lattice constant being small so that for example

$$a\partial_\nu A_\mu(x^\rho) = A_\mu(x^\rho + a\delta_\nu^\rho) - A_\mu(x^\rho), \qquad (3.13)$$

we obtain in this (naïve) continuum limit the action

$$S = \int d^4x\, (-\tfrac{1}{2}\beta e^2\, F_{\mu\nu}(x)\, F^{\mu\nu}(x) - (\kappa e^2/2a^2)\, A_\mu(x^\rho)\, A^\mu(x^\rho)); \qquad (3.14)$$

so naïvely it corresponds to a massive photon model. (Identify $\beta = 1/(2e^2)$ and $m_\gamma^2 = \kappa/(2\beta a^2)$.) Also, there is no gauge symmetry under which the reasonable way would be to put the gauge transformation (22) on the lattice

$$U(\overset{x^\rho}{\bullet}\!\!-\!\!\overset{y^\rho}{\bullet}) \to \Lambda(x^\rho\cdot)\, U(\overset{x^\rho}{\bullet}\!\!-\!\!\overset{y^\rho}{\bullet})\, \Lambda(y^\rho\cdot)^{-1}, \qquad (3.15)$$

because the term $\kappa \sum \mathrm{Re}\, U(\bullet\!\!-\!\!\bullet)$ in the action (3.10) is not invariant under this transformation (3.15). Here in the naïve continuum correspondence

$$\Lambda(x^\rho\cdot) = \exp(ie\lambda(x^\rho)). \qquad (3.16)$$

Only if κ is zero, or at least some specific value, one would expect to find that the photon mass calculated in this model would be zero even when quantum mechanical effects are taken into account. However, according to calculations done by Fradkin & Shenker (1979) and by Banks & Rabinoviči (1979) this is *not* so. Rather there is a whole range of values for the parameters β and κ of the model for which there is a massless photon. It is in fact sufficient that κ is smaller than, and β larger than, some critical values. If we think of the model in analogy with an instrument with buttons by which the values of β and κ are tuned, it is not necessary to fine-tune them to get a zero mass photon (like an automatic frequency control on a radio may make the most accurate fine-tuning superfluous).

The first step in seeing this surprising result is to argue that we can in fact introduce a gauge symmetry into the model in a rather artificial manner.

The trick is to write the model in terms of an unnecessarily large number of field variables: $U_h(\bullet\!\!-\!\!\bullet)$ defined on the links just like the original variables $U(\bullet\!\!-\!\!\bullet)$, and $\Omega_h(\cdot)$ defined on the sites $\cdot$, i.e. the points of the lattice. Both $U_h(\bullet\!\!-\!\!\bullet)$ and $\Omega_h(\cdot)$ are norm unity complex numbers like $U(\bullet\!\!-\!\!\bullet)$. Indeed we write

$$U(\overset{x^\rho}{\bullet}\!\!-\!\!\overset{y^\rho}{\bullet}) = \Omega_h(x^\rho\cdot)\, U_h(\overset{x^\rho}{\bullet}\!\!-\!\!\overset{y^\rho}{\bullet})\, \Omega_h(y^\rho\cdot)^{-1}. \tag{3.17}$$

There are now infinitely many choices of the fields $\Omega_h(\cdot)$ and $U_h(\bullet\!\!-\!\!\bullet)$ which give the same field configuration for the fundamental field $U(\bullet\!\!-\!\!\bullet)$ and thus one may transform around the former without changing the original field variables $U(\bullet\!\!-\!\!\bullet)$. In fact, one has invariance under the formal gauge symmetry

$$\left.\begin{array}{l} U_h(\overset{x^\rho}{\bullet}\!\!-\!\!\overset{y^\rho}{\bullet}) \to \Lambda_h(x^\rho\cdot)\, U_h(\overset{x^\rho}{\bullet}\!\!-\!\!\overset{y^\rho}{\bullet})\, \Lambda_h(y^\rho\cdot)^{-1}, \\ \Omega_h(x^\rho\cdot) \to \Omega_h(x^\rho\cdot)\, \Lambda_h(x^\rho\cdot)^{-1}, \end{array}\right\} \tag{3.18}$$

where $\Lambda_h(\cdot)$ is the gauge function, like $\Lambda(\cdot)$ in (3.15), i.e. a norm unity complex function on the sites. But it is still surprising that a symmetry, which is as (3.18) purely due to a notation with extra many variables, can ensure a physical result, the masslessness of the photon for a whole region, a phase, of (β, κ)-combinations.

Essentially the way this comes about is that the site-defined field $\Omega_h(\cdot)$ has very strong quantum fluctuations when κ is small enough. It can be shown that there are no long range correlations in $\Omega_h(\cdot)$ in this case, so that the $\Omega_h(\cdot)$-field can be ignored as far as low energy or long distance properties of the model are concerned. Then it becomes natural instead of (3.9) to use the identification

$$U_h(\overset{x^\rho}{\bullet}\!\!-\!\!\overset{x^\rho+a\delta_\mu^\rho}{\bullet}) = \exp\,(\mathrm{i}eA_\mu(x^\rho))$$

and one would then have gauge invariance (3.6). The mass term (3.7) will be forbidden and the masslessness of the photon be explained.

How exactly the dynamics of the model work might be complicated to see. The important point is that even with a random choice of (β, κ) there is a finite non-zero probability for obtaining a zero mass photon electrodynamics even with a gauge symmetry, although the latter is from the point of view of the fundamental model introduced just by notation and thus only formal.

This is a consequence of the very reliable estimates of Fradkin & Shenker (1979), Banks & Rabinoviči (1979) and by ourselves (Förster *et al.* 1980) and can be tested on a computer (see Ranft *et al.* 1983).

One may of course ask if we have 'derived' the Maxwell electrodynamics with its zero mass photon if it only appears with finite (non-zero) probability. I think though that one just has to imagine that the fundamental model should consist of several (probably interacting) sets of link fields of the type described and then most likely some of them would produce massless photons. Then we might instead have the problem of why we got only one of them, a question on which we have some premature speculations in a model in which we do not at first include translational invariance.

4. ACHIEVEMENTS OF RANDOM DYNAMICS

Towards the end, let me mention briefly more progress that we have made or are making in the programme on random dynamics: in fact, Antonianidas *et al.* (1983), and Iliopoulos *et al.* (1981) (see also Iliopoulos 1981) have an alternative way of deriving gauge symmetry. They use the renormalization group and find that in a model with no gauge invariance at high energy as one restricts oneself to lower and lower energies and momenta, gauge invariance becomes more and more accurate but never exact. Their model deviates from ours by having no lattice and they are not allowed to introduce a photon mass term but only certain other gauge symmetry breaking terms.

The same method of renormalization group calculations was used by Chadha & Nielsen (1982) and by Ninomiya & Nielsen (1979) to show that for field theories with gauge symmetry already assumed, Lorentz invariance (including rotational invariance) if not valid, is however more accurately satisfied the lower the energy and momentum at which it is studied.

However, I tend to favour an explanation for Lorentz invariance which M. Lehto, M. Ninomiya and I are working on, and which is analogous to the method sketched in the foregoing section. In fact, one may consider gravity as a sort of gauge theory where it is either the Lorentz invariance or translational invariance or both that is gauged. If a theory of gravitation is then obtained without fine-tuning, we will also get Lorentz invariance in the neighbourhood of a space–time point as a side result.

In connection with non-Lorentz invariant models, Chadha and myself (Nielsen 1977, 1978) also find that a space–time with three space and one time dimensions is singled out. A fermion will have much smaller velocities in any further dimensions.

My work with I. Picek (Nielsen & Picek 1982 *a, b*) is more phenomenological seeking possible deviations from Lorentz invariance, i.e. deviations from the principle of relativity, and thus does not strictly speaking belong to the programme of random dynamics. I should like to mention here that there is an experiment by Aronson *et al.* (1982) showing that the parameters of the K^0–$\bar{K}^0$ system in the beam energy range 30–110 GeV are slightly energy dependent, meaning – if taken seriously – that Lorentz invariance is broken. Since it is only a few standard deviations of relativity principle breaking we should be cautious.

Lehto *et al.* (1982) also considered a generalization of electrodynamics with the $A_\mu(x^\rho)$-potential replaced by an antisymmetric tensor 'potential' $A_{\mu\nu}(x^\rho) = A_{\mu\nu}(x^\rho)$, and 'derived' the probable existence of a massless particle analogous to the photon.

In all these investigations quantum mechanics is a very important assumption.

However, S. Chadha, C. Litwin and myself (see Nielsen 1978, 1981) have argued that a random differential equation for the time development of a point in a high dimensional space is likely to approach a fixed point, become approximately linear, and thus be interpreted as the Schrödinger equation. This gives hope of even quantum mechanics resulting from a limit, that in which a long time has elapsed. However, this only resulted because we used as random velocity fields Fourier series with a finite number of random coefficients and these series had a lot of zeros. Taking random differentiable functions, there is only one set of measure zero of motion with a fixed point so that our 'derivation' of quantum mechanics is incorrect with the 'correct' measure (we are grateful to the referee for this comment).

Also it should be mentioned that the discrete symmetries, parity, charge conjugation and time reversal of strong and electromagnetic interactions are well understood in the standard model and that isospin and the old Gell-Mann SU(3) are explicable if one just adds the assumption of the quark masses being small (Weinberg 1979, 1981, 1982; Zee & Wilczek 1979).

Attempting to get details about the standard model N. Brene and I (Brene & Nielsen 1982, 1983) have argued from a model with a random action breaking translational invariance, that gauge groups not having a connected non-trivial centre are likely to break down spontaneously so that the gauge particles (the analogues of the photon) obtain masses. By arguments of this type about what properties of a gauge group endanger it by a breakdown, we tend to favour that at low energies one should find the gauge group $S(U(2) \times U(3))$ (of special i.e. unit determinant 5×5 matrices composed of $U(2)$ and $U(3)$ matrices) which is just the one of the standard model to which one has arrived by more phenomenological considerations.

Froggatt & Nielsen (1979a, 1979b, 1981) have in a random model for mass matrices obtained very inaccurate predictions of, for example the ratio

$$\ln(m_\tau/m_\mu)/\ln(m_\mu/m_e) \approx 0.6 \pm 0.5,$$

where the masses of the three lightest charged leptons are denoted m_e (electron), m_μ (muon), and m_τ (tau lepton). It agrees almost too well with experiment.

Recently we have been speculating on how to obtain geometry out of a (random) gauge theory model. Fu Ying Kai and I found that in an anisotropic lattice electrodynamics it is possible to produce what we call a layered phase. This is a state in which a charged particle would be able to move only along some layers but not across them. The idea is then that it is the layers (or one layer curled up) that make up the four dimensional geometry we know, the geometry outside being unimportant. Also we hope then to 'derive' a principle of locality which says that there is no direct interaction except over very small distances: no action at long distances. This also looks promising.

5. Concluding remarks

It seems that there are indeed some ways of deriving many of the most typical features of what we know today from more random fundamental laws of nature, e.g. gauge symmetry, Lorentz invariance, linearity of the Schrödinger equation (in jeopardy though), discrete symmetries, and $(3+1)$-dimensionality space–time. We may even get some information on which gauge group to expect and some very crude information on lepton and quark masses, and Cabibbo angles.

The picture of random dynamics – which we find somewhat promising – may be said to be roughly analogous to the development of species of animals and vegetation throughout the history of the Earth. Also – for me at least – the Darwinian ideas of development of species have provided inspiration for the presently described work.

The analogy should be this: the geological time for the development of life corresponds to the logarithm of the length scale (or to minus the logarithm of the energy and momentum scale) for physics. The various geological periods may correspond to various branches of physics, many of which are probably yet to be discovered. The idea of random dynamics that various features – symmetries, linearity properties, etc. – arise at various levels corresponds to the various inventions or development of organs or mechanisms such as the RNA amino acid code and production system, muscles, brain, and social behaviour. Maybe there is even a correspondence in that the development of life out of non-living chemistry (and physics) is as especially speculative as the development of physics out of chaotic fundamental laws of nature may also seem.

It is a pleasure to thank all my collaborators whose work I have presented and Zheng Hai-Bin for finding the reference to Ranft (1983). P. Mansfield is thanked for help in improving the English.

References

Antonianidas, I., Iliopoulos, J. & Tamaros, T. N. 1983 *On the infrared stability of gauge theories.* LPTENS 83/13.
Aronson, S. H., Bock, G. J., Cheng, Hai-Yang & Fischbach, E. 1982 *Phys. Rev. Lett.* **48**, 1306.
Banks, T. & Rabinoviči, L. 1979 *Nucl. Phys.* B **160**, 349.
Bjorken, J. & Drell, S. 1964 *Relativistic quantum mechanics*, Appendix A. New York: McGraw-Hill.
Brene, N. & Nielsen, H. B. 1982 *Why the standard model group should have a connected centre.* Niels Bohr Institute preprint NBI-HE-82-29.
Brene, N. & Nielsen, H. B. 1983 *Standard model group – survival of the fittest.* Niels Bohr Institute preprint NBI-HE-83-04. (Submitted to *Nucl. Phys.* B.)
Chadha, S. & Nielsen, H. B. 1982 *Lorentz invariance as a low energy phenomenon.* Niels Bohr Institute preprint NBI-HE-82-42.
Eddington, A. 1946 *Fundamental theory.* Cambridge University Press.
Fradkin, E. & Shenker, S. 1979 *Phys. Rev.* D **19**, 3682.
Froggatt, C. D. & Nielsen, H. B. 1979a *Nucl. Phys.* B **147**, 277.
Froggatt, C. D. & Nielsen, H. B. 1979b *Nucl. Phys.* B **164**, 114.
Froggatt, C. D. & Nielsen, H. B. 1981 *Phys. Lett.* B **106**, 487.
Förster, D., Ninomiya, M. & Nielsen, H. B. 1980 *Phys. Lett.* B **94**, 135.
Iliopoulos, J. 1981 Unification. In *Particle physics* 1980 (ed. J. Andrić, J. Dadic & N. Zovko), p. 1. Amsterdam: North-Holland.
Iliopoulos, J., Nanapoulos, D. V. & Tamaros, T. N. 1980 *Phys. Lett.* B **94**, 141.
Lehto, M., Ninomiya, M. & Nielsen, H. B. 1982 *Phys. Lett.* B **115**, 129.
Nielsen, H. B. 1977 In *Fundamentals of quark models* (ed. J. M. Barbour & A. T. Davies), p. 528. Glasgow: University of Glasgow.
Nielsen, H. B. 1978a *Gamma* **36**, 3.
Nielsen, H. B. 1978b *Gamma* **37**, 35.
Nielsen, H. B. 1981 In *Particle physics* 1980 (ed. I. Andrić, I. Dadic & N. Zovko), p. 125. Amsterdam: North-Holland.
Nielsen, H. B. 1983 In *Random dynamics* (Proceedings of Arctic Physics Summer School, Äkäslompo, Finland). Berlin: Springer.
Nielsen, H. B. & Picek, I. 1982a *Phys. Lett.* B **114**, 141.
Nielsen, H. B. & Picek, I. 1982b *Nucl. Phys.* B **141**, 153.
Ninomiya, M. & Nielsen, H. B. 1979 *Nucl. Phys.* B **141**, 153.
Ranft, F., Kripfganz, F. & Ranft, G. 1983 *Phys. Rev.* D **28**, 360.
Shenker, S. 1980 Symmetry breaking operators in gauge theories. Research plan sent to Department of Energy.
Slater, N. B. 1957 *The development and meaning of Eddington's 'fundamental theory'.* Cambridge University Press.
Weinberg, S. 1979a *Phys. Rev. Lett.* **43**, 1566.
Weinberg, S. 1979b *Phys. Rev. Lett.* **43**, 1571.
Weinberg, S. 1981 In *The second workshop on grand unification* (ed. J. P. Leveille, L. R. Sulak & D. G. Unger), p. 297. Basel: Birkhauser.
Weinberg, S. 1982 *Phys. Rev.* D **26**, 287.
Zee, A. & Wilczek, F. 1979 *Phys. Rev. Lett.* **43**, 1571.

Discussion

J. G. TAYLOR (*Department of Mathematics, King's College London, Strand, London WC2R 2LS, U.K.*). There seems to be considerable arbitrariness in Dr Nielsen's talk in spite of his claim to obtain much from few assumptions. In particular he uses a regular lattice, with only one lattice spacing. A more general approach would be to use a lattice with many spacings, or even a random lattice. Yet in the latter case recent work by Saclay (C. Itzykson and his collaborators) has shown that for the free field (in one dimension) the spectrum of states does not have a natural cut-off. Nor is the problem of fermions easy to resolve; recent questions on the degeneracy of massless fermions on the lattice are apparently completely open on such lattices. Surely these problems have to be satisfactorily analysed before any claims to having constructed 'random dynamics' in a general way can be substantiated.

H. B. NIELSEN. I thank Professor Taylor for calling my attention to work by C. Itzykson, T. D. Lee *et al.* He is certainly right that the lattice electrodynamics model without gauge symmetry which I described has several arbitrary features such as the lattice being a regular hypercubic one. The excuse for this is that it is one of the basic ideas in the project of random dynamics that the detailed features of a model are not essential for what results in the limit of, say, low energy, to which we have experimental access today, so we hope that the arbitrary features are not essential.

For instance it is almost certainly not important whether the lattice is a simple cubic one or whether we would take some more complicated but still regular lattice structure. That we took the lattice to be a regular repetition of the same unit is an assumption justified as a mild form of the principle of translational invariance. That makes it less arbitrary, but of course, to complete the random dynamics project we should then also provide a 'derivation' of translational invariance. This M. Lehto, M. Ninomiya and myself have attempted to do in a model having a lattice the structure of which is dynamical and therefore normally highly irregular due to quantum fluctuations.

It is correct that the problem of putting fermions on a lattice when they belong to a parity non-invariant system of representations is a severe difficulty for random dynamics. This difficulty with chiral fermions on a lattice is especially a problem for our understanding of gauge invariance (the work which I treated in some detail). The reason for this being so severe a problem is (1) that we need a cutoff to be taken seriously, preferably a lattice, and (2) we obtain gauge invariance by definition, and therefore exactly, even at the lattice scale. With these requirements we have no-go theorems (by M. Ninomiya and myself) that seem to leave no way open for describing the phenomenology of parity violation by the weak interactions without giving up other principles, such as locality, in a drastic manner.

The best way out might be to allow the quarks and leptons to be bound states of some system of fermions belonging to representations with equally many species of right and left handed Weyl particles. Maybe also the openness of the fermion problem on the Itzykson–Lee type irregular lattice – which Professor Taylor mentions – gives some hope that species doubling might be avoided.

The problem of species doublers is presumably a general problem for any claim of taking an ultraviolet cut-off seriously, and especially for us.

Einstein gravitation as a long wavelength effective field theory

By S. L. Adler

The Institute for Advanced Study, Princeton, New Jersey 08540, U.S.A.

A possible resolution of the difficulties in quantizing general relativity is provided by the suggestion that Einstein gravitation is not a fundamental field theory, but rather is a long-wavelength effective field theory, arising as a scale-symmetry-breaking effect in a renormalizable fundamental theory. In unified theories of this type, Newton's constant will be calculable in terms of fundamental particle masses. The history and current status of these ideas is reviewed.

In the conventional picture of the fundamental forces of physics, as recently reviewed in Weinberg (1980), gravitation appears on a quite different footing from the weak, strong and electromagnetic interactions of the matter fields. The total dynamics, in the usual formulation, is governed by an action functional

$$S = \int d^4x (-g)^{\frac{1}{2}} (\mathscr{L}_m + \mathscr{L}_{grav}). \tag{1a}$$

Here $\mathscr{L}_m$ is a renormalizable matter Lagrangian density, containing only dimensionless coupling constants, and $\mathscr{L}_{grav}$ is the Einstein–Hilbert gravitational Lagrangian

$$\mathscr{L}_{grav} = (1/16\pi G) R, \tag{1b}$$

with R the scalar curvature. Since the coupling constant $(16\pi G)^{-1}$ appearing in the gravitational action has the dimensionality $(mass)^2$, quantization of the gravitational part of (1) leads to a nonrenormalizable field theory. Furthermore, in the conventional view, there is no mechanism for relating the gravitational mass scale set by $G^{-\frac{1}{2}}$ to the unification mass of the matter fields. Gravitation thus appears as a phenomenon quite outside the usual framework of theoretical ideas on which elementary particle theory is based.

This statement of the problem of 'quantizing gravitation' assumes, however, that the Einstein–Hilbert action is the fundamental quantum action for gravitation. Since all gravitational experiments done to date involve very long wavelengths ($\lambda \gtrsim 10$ cm), there is in fact no experimental evidence for this assumption. Thus, before proceeding to study quantum gravity, we must address the question: is the Einstein theory a fundamental theory, or is it a long wavelength effective field theory?

A familiar example of a long wavelength effective field theory is provided by the Fermi theory of weak interactions, as extended by investigations in particle physics over the last fifteen years. For energies well below 100 GeV (i.e. for wavelengths much longer than 10^{-16} cm), the weak interactions are described by the current–current effective action

$$S_{eff}[(\text{fermions})] = \int d^4x (\mathscr{L}_{eff}^{ch} + \mathscr{L}_{eff}^{n}), \tag{2}$$

where

$$\mathscr{L}_{eff}^{ch} = 2^{-\frac{1}{2}} G_F (j_{ch}^\lambda + J_{ch}^\lambda)(j_{ch\lambda}^\dagger + J_{ch\lambda}^\dagger),$$

$$j_{ch}^\lambda = \bar{e}\gamma^\lambda (1-\gamma_5) \nu_e + \mu, \tau \text{ terms},$$

$$J_{ch}^\lambda = \bar{u}\gamma^\lambda (1-\gamma_5)(\text{d}\cos\theta_C + \text{s}\sin\theta_C) + \text{c, t, b terms, etc.},$$

with G_F the dimensional constant

$$G_F \approx 10^{-5}/m_p^2, \quad m_p \text{ is the proton mass.} \tag{3}$$

As expected for a theory with a dimensional coupling constant, the Fermi theory is non-renormalizable, and repeated attempts to quantize the weak interactions starting from the Fermi theory as the fundamental quantum action have met with frustration. It is now known that the Fermi theory is only a long-wavelength effective theory for the weak interactions. The fundamental quantum theory for the weak (and electromagnetic) interactions is the renormalizable gauge theory of Glashow, Salam and Weinberg (G.S.W), in which the weak interactions are mediated by the exchange of massive intermediate vector bosons, which obtain their masses from a symmetry-breaking mechanism involving Higgs scalar bosons. When all fermion energies are much lower than the intermediate boson masses, the Fermi theory is recovered from the G.S.W. theory as a low-energy, long-wavelength effective theory for the fermions. In the language of functional integrals, the relation between the Fermi effective theory and the G.S.W. fundamental theory is given by

$$e^{iS_{\text{eff}}[(\text{fermions})]} = \int d(\text{bosons}) \, e^{iS_{\text{fund}}[(\text{fermions}),(\text{bosons})]}, \tag{4}$$

and from (4) one readily finds an experimentally verified formula relating the Fermi constant to the parameters of the fundamental theory,

$$2^{-\frac{1}{2}} G_F = e^2/8 m_W^2 \sin^2 \theta_W, \tag{5}$$

where

$$e = \text{electric charge},$$

$$m_W = \text{charged intermediate boson mass},$$

$$\theta_W = \text{SU}(2) - \text{U}(1) \text{ mixing angle}.$$

Returning now to gravity, let us assume that the strategy that has worked so successfully for the weak interactions should also be applied to the problem of quantizing gravitation. Thus, we shall assume that the fundamental gravitational action is the renormalizable and classically scale-invariant action

$$S_{\text{fund}} = \int d^4x (-g)^{\frac{1}{2}} (\alpha C_{\mu\nu\lambda\sigma} C^{\mu\nu\lambda\sigma} + \beta R^2), \tag{6}$$

with $C_{\mu\nu\lambda\sigma}$ the Weyl tensor (the traceless part of the Riemann curvature tensor $R_{\mu\nu\lambda\sigma}$). Quantum corrections break the scale symmetry of (6), and as a result induce an Einstein–Hilbert effective action in the low-energy, long-wavelength limit; this effective action governs observed gravitational phenomena (just as the Fermi effective theory describes low energy β-decay physics) but is not the fundamental quantum field theory action. The 'induced gravitation' approach just sketched has been actively studied over the last few years, as reviewed in Adler (1982), and is the viewpoint adopted in the remainder of our discussion here.

Continuing to develop the analogy between the Fermi and the Einstein–Hilbert Lagrangians, let us ask what characteristic mass appears in the coupling constant for each effective action. In the microscopic units where $\hbar = c = 1$, the action S is dimensionless, the integration measure d^4x has dimension (mass)$^{-4}$, and thus the Lagrangian density $\mathscr{L}$ has dimension (mass)^4. From this fact, and the dimensionalities of the fields appearing in the effective actions, we infer the dimensionalities of the coupling constants already quoted above:

Fermi theory:

$$S_{\text{eff}} = \int \mathrm{d}^4x\, 2^{-\frac{1}{2}} G_F \underbrace{\overbrace{\bar\psi \ldots \psi \bar\psi \ldots \psi}^{\text{dimension (mass)}^4}}_{\text{dimension (mass)}^6}, \quad \psi = \text{fermion field},$$

$$\Rightarrow G_F \text{ has dimension (mass)}^{-2}; \tag{7a}$$

Einstein–Hilbert theory:

$$S_{\text{eff}} = \int \mathrm{d}^4x\, (-g)^{\frac{1}{2}} \frac{1}{16\pi G} \underbrace{\overbrace{R}^{\text{dimension (mass)}^4}}_{\text{dimension (mass)}^2}$$

$$\Rightarrow (16\pi G)^{-1} \text{ has dimension (mass)}^2. \tag{7b}$$

Comparing (7a) with (7b), we see that there is an important difference between the behaviour of the coupling constant in the two cases. Since the Fermi coupling G_F has the dimensionality of mass to a *negative* power, the dominant contributions to G_F come from the *smallest* mass intermediate states with the relevant quantum numbers, which are the intermediate bosons with mass $m_W \approx 80 \text{ GeV}$. By contrast, the inverse Newton's constant G^{-1} has the dimensionality of mass to a *positive* power, and hence the dominant contributions to G^{-1} will come from the *largest* characteristic mass scale appearing in physics. This is presumably the Planck mass M_P of order 10^{19} GeV, where the gravitational interactions of elementary particles become of comparable importance to their electro-weak and strong interactions.

Just as G_F can be calculated in terms of the more fundamental parameters of the G.W.S. gauge theory, in induced gravity theories one expects Newton's constant G to be calculable in terms of fundamental particle masses. To see how this can come about, we draw on the fact that when $\alpha = g^{-2} > 0$, $\beta = g'^{-2} > 0$, the action of (6) leads to an *asymptotically free* quantum theory, just as the current candidates for unified matter theories are also asymptotically free quantum theories. The term 'asymptotically free' refers to the behaviour of the coupling constant which, when radiative corrections are included, is changed from a true constant to a running function of the dominant dimensional variable, assumed for simplicity in the following discussion to be an energy E. In other words, in an asymptotically free theory, quantum effects lead to the replacement

$$g^2 \to g^2_{\text{run}} = \frac{g^2(\mu)}{1 + bg^2(\mu)\ln(E/\mu)}, \tag{8}$$

with μ an arbitrary reference mass and with $b(>0)$ a constant characteristic of the theory. As the energy E approaches infinity, (8) implies

$$g^2_{\text{run}} \xrightarrow[E\to\infty]{} 0, \tag{9}$$

which leads to the vanishing of forces, and hence to free field theory behaviour, in the asymptotic high energy limit. On the other hand, the physics described by (8) becomes strongly interacting at low energies, as is readily seen by rewriting (8) in the form

$$g^2_{\text{run}} = \frac{1}{b\ln(E/\mathcal{M})}. \tag{10}$$

Here

$$\mathcal{M} = \mu\, \mathrm{e}^{-1/bg^2(\mu)}, \tag{11}$$

[65]

is a mass parameter, which can be shown to be μ-independent, and which characterizes the theory in the sense that E of order $\mathcal{M}$ defines the strong coupling régime. We see that as a result of including radiative corrections, a one parameter family of classical theories characterized by their values of the dimensionless coupling g^2, has been replaced by a one parameter family of quantum theories characterized by their values of the dimension (mass)^1 scale mass $\mathcal{M}$. As a result of this phenomenon, called *dimensional transmutation*, all dimensional physical parameters in an asymptotically free theory are calculable in terms of $\mathcal{M}$, much as all radiative effects in the familiar, non-asymptotically free, case of quantum electrodynamics are calculable in terms of the fine structure constant α.

Let us now suppose that the gravitational theory of (6) and the unified matter theories, both of which are asymptotically free, can be further unified into an asymptotically free theory with a single classical coupling constant g. Then as a result of dimensional transmutation, g is replaced as a parameter in the quantized theory by a mass parameter $\mathcal{M}$, which presumably should be identified with M_P. All particle masses and G^{-1} will then be calculable in terms of $\mathcal{M}$, or eliminating $\mathcal{M}$, the ratio

$$G^{-1}/(\text{any particle mass})^2, \tag{12}$$

will be calculable. This scenario naturally accommodates the fact that G^{-1} is of order M_P^2, but does not explain either why the cosmological constant is very small, or why elementary particle masses are small, on the scale of the Planck mass. These unexplained features are a problem in *all* unified models to date, and presumably will eventually be explained by specific kinematical or dynamical features of the ultimate unifying theory.

Having sketched the principal qualitative features of induced gravity theories, let me now survey in a somewhat more technical way their history and current status.

(1) The suggestion that the Einstein–Hilbert theory is an effective field theory was first made by Sakharov (1967), who proposed that the Einstein–Hilbert action arises from the quantum fluctuations of quantized matter fields in a curved background manifold. (For a survey of this and related early work, see ter Haar *et al.* (1982) and Adler (1982).)

(2) Subsequently, a number of authors (for full references, see Adler 1982) studied Higgs-type models in a curved background manifold, with the action

$$S = \int d^4x (-g)^{\frac{1}{2}} [\tfrac{1}{2}\epsilon\phi^2 R - V(\phi) + ...]. \tag{13a}$$

In (13), $V(\phi)$ is a symmetry-breaking double well potential with a global minimum at $\phi^2 = \bar{\phi}^2$, so that when quantized around the stable vacuum (13a) yields an Einstein–Hilbert action, with the gravitational constant given by

$$1/16\pi G = \tfrac{1}{2}\epsilon\bar{\phi}^2. \tag{13b}$$

In models of this type, the renormalized coupling ϵ is necessarily an independent parameter, and so G^{-1} is not calculable in terms of fundamental mass parameters.

(3) Further progress stemmed from the observation (Adler 1980a) that in *scalar-free* theories with dynamical breaking of scale invariance, an Einstein–Hilbert action is induced with a calculable Newton's constant G^{-1}. Since a pure non-Abelian gauge theory satisfies these calculability criteria, the simplest model for induced gravity is thus a non-Abelian gauge theory (g.t.) quantized on a classical background manifold, for which the analogue of (4) reads

$$e^{iS_{\text{eff}}[g_{\mu\nu}]} = \int d[A_\lambda] e^{iS_{\text{g.t.}}[A_\lambda, g_{\mu\nu}]}. \tag{14}$$

(4) In the background metric model, the coefficients of the various terms in the long-wavelength expansion of S_{eff},

$$S_{\text{eff}}[g_{\mu\nu}] = \int d^4x (-g)^{\frac{1}{2}} \left[\frac{1}{16\pi G_1}(R - 2\Lambda_1) + \alpha C_{\mu\nu\lambda\sigma}C^{\mu\nu\lambda\sigma} + \beta R^2 + \ldots \right], \tag{15}$$

can be extracted by taking successive metric variations. Acting on the left and right sides of (14) with $g_{\mu\nu}\delta/\delta g_{\mu\nu}$ and specializing to flat space–time gives the usual formula for the induced cosmological constant,

$$-(2\pi)^{-1}\Lambda_1/G_1 = \langle T^\mu_\mu \rangle_0, \tag{16}$$

where T^μ_μ is the trace of the gauge theory stress-energy tensor and where $\langle \rangle_0$ denotes the flat space–time vacuum expectation value. Acting with $(g_{\mu\nu}\delta/\delta g_{\mu\nu})^2$ gives a formula (Adler 1980b, Zee 1981) for the induced gravitational constant,

$$\left. \begin{aligned} \frac{1}{16\pi G_1} &= \frac{-\mathrm{i}}{96}\int d^4x\, x^2 \langle \mathcal{T}(\tilde{T}(x)\,\tilde{T}(0))\rangle_0, \\ \tilde{T}(x) &= T(x) - \langle T(x)\rangle_0,\ T \equiv T^\mu_\mu, \end{aligned} \right\} \tag{17}$$

and when higher derivative terms are retained, also a formula (Zee 1982) for the coefficient of the induced R^2 term,

$$\beta = \frac{-\mathrm{i}}{13\,824}\int d^4x\,(x^2)^2 \langle \mathcal{T}(\tilde{T}(x)\,\tilde{T}(0))\rangle_0. \tag{18}$$

For a general pure non-Abelian gauge theory, the coefficients Λ_1/G_1, G_1^{-1} and β given by (16)–(18) are all calculable in terms of the gauge theory scale mass $\mathcal{M}$.

(5) To go beyond the background metric model, one must take into account the fact that in a realistic theory, gravity (i.e. the metric $g_{\mu\nu}$) is also quantized. To do this, we make the decomposition

$$g_{\mu\nu} = \bar{g}_{\mu\nu} + h_{\mu\nu}, \tag{19}$$

where $\bar{g}_{\mu\nu}$ is the average background metric and where $h_{\mu\nu}$ is a quantum fluctuation. The functional integral formalism for background field quantization then implies that $\bar{g}_{\mu\nu}$ is self-consistently determined by a classical variational principle,

$$\delta S_{\text{eff}}[(\bar{\phi}_m), \bar{g}_{\mu\nu}] = 0, \tag{20}$$

with $\bar{\phi}_m$ the average matter fields. In the long-wavelength limit, (20) yields the classical Einstein equations. The self-consistent structure of the calculation is reflected in the fact that the functional form of S_m is itself determined by the quantum fluctuations $h_{\mu\nu}$ around the mean value $\bar{g}_{\mu\nu}$. Acting with $(\bar{g}_{\mu\nu}\delta/\delta\bar{g}_{\mu\nu})^2$ as described above, one obtains a formal functional integral expression for the induced gravitational constant G_1^{-1} including quantum gravity effects (Adler 1982).

(6) Finally, let us return to the basic question: what is the fundamental gravitational action? In the absence of additional non-metric degrees of freedom (which may well be present!), the renormalizable candidates are

$$\alpha C_{\mu\nu\lambda\sigma}C^{\mu\nu\lambda\sigma} + \beta R^2, \tag{21a}$$

which is scale invariant, and

$$\alpha C_{\mu\nu\lambda\sigma}C^{\mu\nu\lambda\sigma}, \tag{21b}$$

which is conformally invariant. If the Lagrangian of (21b) leads to a finite induced R^2 term (as in the background metric model calculation of (18)) then it is a viable candidate for the fundamental action, while if the induced R^2 term arising from (21b) is divergent, then an R^2 counter term is needed, as in (21 a). The classic objection to the fourth-order Lagrangians of (21) is that, in a small fluctuation analysis, they have an energy spectrum which is unbounded from below. In this connection there is a very interesting new global result (Boulware et al. 1983): *for the Lagrangian (21b), as well as for (21a) with $\alpha\beta > 0$, all exact classical solutions have zero energy.* Hence these Lagrangians may well lead in a natural way to satisfactory quantum field theories (without recourse to such unappealing devices as negative metric quantization and special integration contour prescriptions). Thus at this point, the whole subject of the quantization of fourth-order gravitational theories has been re-opened, and is an exciting direction for future research in quantum gravitation.

This work was supported by the U.S. Department of Energy under grant no. DE-ACO2-76ERO2220.

References

Adler, S. L. 1980a *Phys. Rev. Lett.* **44**, 1567–1569.
Adler, S. L. 1980b *Physics Lett.* B **95**, 241–243.
Adler, S. L. 1982 *Rev. mod. Phys.* **54**, 729–766.
Boulware, D. G., Horowitz, G. T. & Strominger, A. 1983 *Phys. Rev. Lett.* **50**, 1726–1729.
ter Haar, D., Chudnovsky, D. V. & Chudnovsky, G. V. (ed.) 1982 *A. D. Sakharov collected scientific works.* New York and Basel: Marcel Dekker.
Sakharov, A. D. 1967 *Dokl. Akad. Nauk. SSSR* **177**, 70–71. (*Soviet Phys. Dokl.* **12**, 1040–1041 (1968)).
Weinberg, S. 1980 *Rev. mod. Phys.* **52**, 515–523.
Zee, A. 1981 *Phys. Rev.* D **23**, 858–866.
Zee, A. 1982 *Physics Lett.* B **109**, 183–186.

Unification and supersymmetry

By J. Ellis[†]
CERN, Theory Division, CH-1211 Geneva 23, Switzerland

Current attempts to construct unified theories of fundamental particles and their interactions are described, with emphasis on their ability to understand the values of the fundamental constants. Examples include grand unified theories, which enable one to estimate the fine structure constant, the neutral weak interaction mixing parameter and certain quark masses. Finally, a review will be presented of the prospects offered by supersymmetry for understanding the scale of the weak interactions and for an eventual unification with gravity.

Introduction

My plan in this talk is to lead you further down the primrose path of unification to which you were introduced by Llewellyn Smith (1983) and by Weinberg (1983). Starting from the basic ideas of grand unification which you have already met, we will go on to more recent approaches to unification such as technicolour (Farhi & Susskind 1981), supersymmetry (Fayet & Ferrara 1977) and supergravity (van Nieuwenhuizen 1981). The emphasis throughout will be on ideas for understanding the values of the apparently 'fundamental' constants.

In §1 we will count the parameters of the standard $SU(3) \times SU(2) \times U(1)$ model and find that there are at least 20 'fundamental' constants to be explained. The philosophy of conventional grand unification (Georgi & Glashow 1974; Georgi *et al.* 1974) is reviewed in §2, and we find (Ellis & Nanopoulos 1981) that it is consistent only if the fine structure constant lies in the range $\frac{1}{170} < \alpha < \frac{1}{120}$. Section 3 introduces simple models for grand unification and we find that they enable us to calculate successfully the ratios of some of the standard model parameters. For example, charge quantization $|Q_e/Q_p| = 1$ is explained and the neutral weak mixing angle θ_W is calculable (Georgi *et al.* 1974), as well as some quark masses which are related to charged lepton masses (Chanowitz *et al.* 1977; Buras *et al.* 1978). These sections are brief, since many reviews of classical grand unification exist and the topics have already been touched on at this meeting (Llewellyn Smith 1983; Weinberg 1983). In §4 we will assess critically the significance for grand unification of the recent negative results (Bionta *et al.* 1983; Goldhaber 1983) of a search for baryon decay. Recent calculations (Brodsky *et al* 1983; Isgur & Llewellyn Smith 1983) relating the baryon decay rate to the short-distance behaviour of the proton form factor suggest that the baryon decay amplitude may be rather larger (for a given value of m_X) than previous SU(6) estimates (Isgur & Wise 1982) had suggested, which would deepen the apparent conflict between experiment and conventional minimal grand unified theories (GUTs). However, even before the negative results of this search, theorists had grown dissatisfied with minimal GUTs with their 21 'fundamental' parameters, which do not represent a significant decrease from the 20 'fundamental' parameters of the standard $SU(3) \times SU(2) \times U(1)$ model! Establishing and maintaining a small weak interaction scale (m_W/M_P of order 10^{-17}) is a severe difficulty for conventional GUTs

[†] On leave of absence at SLAC, P.O. Box 4349, Stanford, California 94305, U.S.A.

with elementary scalar Higgs fields. Attempts to understand the weak interaction scale (the 'hierarchy problem') (Gildener & Weinberg 1976; Gildener 1976) are described in §5. They include dynamical symmetry breaking, called here technicolour (Farhi & Susskind 1981), which renders m_W calculable, but must be supplemented by epicycles if one is to get non-zero quark and lepton masses. This complication is the source of severe phenomenological difficulties (Dimopoulos & Ellis 1981) which have not yet been overcome. A currently favoured alternative strategy for understanding the weak interaction scale is supersymmetry (Fayet & Ferrara 1977). It seems that one needs local supersymmetry (supergravity) for building realistic models (Alvarez-Gaumé et al. 1983; Ibáñez & López 1983; Ellis et al. 1983a) So far, these models tend to have m_W of order the gravitino mass, although there are models (Ellis et al. 1983a) in which the weak interaction scale is fixed dynamically by an analogue of dimensional transmutation (Coleman & Weinberg 1973). Finally, in §6 we mention some possible strategies (Ellis et al. 1979; Ellis et al. 1983a) for understanding the grand unification scale m_X and the magnitude of the gauge coupling at energies of order m_X. Ultimately one may hope (Ellis et al. 1980c) that all the 'fundamental' constants may be calculable using an underlying $N = 8$ supergravity theory, but so far this is a dream beyond our dynamical understanding.

1. The parameters of the standard model

As Llewellyn Smith (1983) has told you, the standard model of strong, weak and electromagnetic interactions is based on the gauge group $SU(3) \times SU(2) \times U(1)$ and contains three generations of quarks and leptons (u, d, e, ν_e), (c, s, μ, ν_μ) and (t, b, τ, ν_τ). We have three independent gauge couplings g_3, g_2 and g_1 for the three factors of the gauge group. The fine structure constant

$$\alpha = (g_2^2/4\pi) \sin^2 \theta_W, \tag{1}$$

where θ_W describes mixing between the neutral $SU(2)$ and $U(1)$ currents:

$$\sin^2 \theta_W = \tfrac{3}{5} g_1^2 / (g_2^2 + \tfrac{3}{5} g_1^2). \tag{2}$$

You have heard that QCD with massless quarks has no free parameters: this is true as long as there is no external scale on which to measure g_3. We will take the Planck mass

$$M_P \equiv G_n^{-\frac{1}{2}} \approx 1.2 \times 10^{19} \text{ GeV}, \tag{3}$$

as our fundamental physical scale. It is then meaningful to ask what is the value of g_3 at a specified energy scale, say $E = 10^{-18} M_P$. The gauge interactions of $SU(3) \times SU(2) \times U(1)$ also require for their specification two non-perturbative CP-violating vacuum parameters θ_3, θ_2. The QCD angle θ_3 is in principle observable via the neutron electric dipole moment

$$d_n \approx 3 \times 10^{-16} \theta_3 \, e \, \text{cm}, \tag{4}$$

and from the present experimental upper limit (Altarev et al. 1981) on d_n we know that

$$\theta_3 < 2 \times 10^{-9}. \tag{5}$$

Understanding the smallness of θ_3 is a major theoretical puzzle. The $SU(2)$ vacuum angle θ_2 is practically unobservable because non-perturbative weak interactions are negligible: in principle θ_2 could be $O(1)$, though this seems highly implausible. Leaving the gauge sector we encounter six quark masses and three lepton masses as 'fundamental' parameters. In the standard model these are derived from underlying Yukawa interactions along with the three charged weak interaction mixing angles which generalize the 4-quark Cabibbo mixing angle to the case of six quarks, and a single CP-violating phase δ (Kobayashi & Maskawa 1973) which is blamed for the observed CP-violation in K^0 decays. Finally, in the standard model there are two weak boson masses to be specified: $m_{W^\pm}$ and the Higgs boson mass m_H. The strength of the Fermi weak interaction is derived from the 'fundamental' parameters g_2 and m_W:

$$G_F/2^{\frac{1}{2}} = g_2^2/8m_W^2 + \text{(radiative corrections)}. \tag{6}$$

TABLE 1. PARAMETER COUNTS

standard $SU(3) \times SU(2) \times U(1)$ model	minimal $SU(5)$ GUT	type
3: g_3, g_2, g_1	1: g_5	gauge couplings
2: θ_3, θ_2	1: θ_5	vacuum angles
6	6	quark masses
3	0	lepton masses
3: θ_i	3: θ_i	charged weak mixing angles
1: δ	3: δ_j	CP-violating phases
2: m_W, m_H	7	boson masses
20	21	total

The above parameters ('fundamental' constants) are listed in table 1: there are a total of 20 of them. However, even this list is incomplete as we have assumed the standard electroweak quantum numbers for quarks, leptons and Higgs fields. We do not know why left-handed fermions should sit in doublets of $SU(2)$ while right-handed fermions should be singlets of $SU(2)$. Another egregious mystery is the choice of weak $U(1)$ hypercharges which have arranged themselves so as to respect charge quantization:

$$|Q_e/Q_p| = 1 + O(10^{-20}). \tag{7}$$

There is no explanation for this quantization within the standard model.

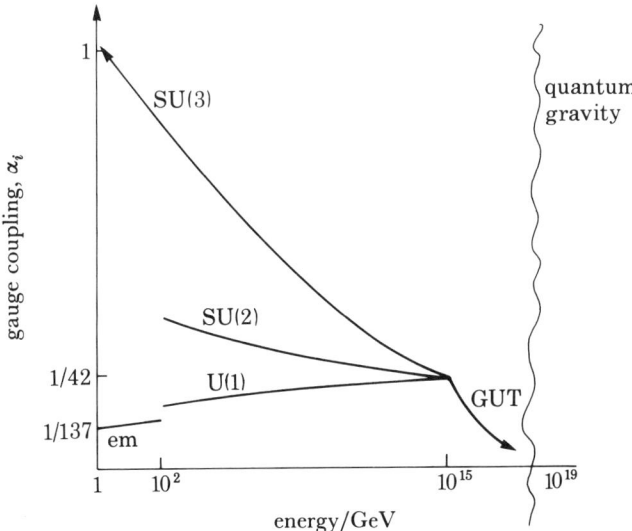

FIGURE 1. The approach of the $SU(3) \times SU(2) \times U(1)$ gauge couplings to be unified at a scale $m_X = O(10^{15})$ GeV well below the Planck mass of 10^{19} GeV at which quantum gravity effects become important.

2. The philosophy of grand unification

This is indicated in figure 1 and has already been explained to you by Llewellyn Smith (1983). Given the absurd assumption that there is a grand desert with no prior oases of new particle or interaction thresholds, asymptotic freedom drives the strong SU(3) coupling down to meet the SU(2) and U(1) weak couplings at an energy scale of order 10^{15} GeV, to be identified with the masses of superheavy gauge vector bosons m_X. This unification scale is astronomically high because of the logarithmically slow evolution of the gauge couplings:

$$m_X = m_B \exp\left(O(1)/\alpha\right). \tag{8}$$

It is important for the consistency of the whole GUT philosophy that m_X be less than 10^{19} GeV, so that the neglect of gravity is a reasonable first approximation, while m_X must be larger than about 10^{14} GeV if baryons are not to decay more rapidly than experiment allows. For m_X (equation (8)) to lie within this range, we must have (Ellis & Nanopoulos 1981)

$$\tfrac{1}{170} < \alpha < \tfrac{1}{120} \tag{9}$$

as a consistency condition for the GUT philosophy. Happily enough, $\alpha = \tfrac{1}{137}$ lies within the range (equation (9)), and we can go on to look at specific GUT models.

3. Simple models

We must look for simple non-Abelian groups of rank $R \geqslant 4$ in order to be able to include the $SU(3) \times SU(2) \times U(1)$ group of the standard model. Georgi & Glashow (1974) found that the only acceptable group of rank 4 was SU(5) which contains 24 gauge bosons. Half of these are the familiar photon, eight gluons, recently discovered $W^{\pm}$ (Arnison *et al.* 1983 a; Banner *et al.* 1983) and the even more recent Z^0 (Arnison *et al.* 1983 b). Then there are 12 superheavy gauge bosons X and Y which will mediate new hyperweak interactions violating baryon number, B, conservation. The gauge bosons mediate interactions between three generations of fermions, each of which contains 15 helicity states assigned unaesthetically to a reducible $\bar{5} + \underline{10}$ representation of SU(5). The lightest $\bar{5}$ representation is

$$\bar{5} = \begin{pmatrix} \bar{d}_R \\ \bar{d}_Y \\ \bar{d}_B \\ e^- \\ \nu_e \end{pmatrix} \left. \begin{matrix} \\ \\ \\ \end{matrix} \right\} \text{strong SU(3)} \atop \left. \begin{matrix} \\ \end{matrix} \right\} \text{weak SU(2)} \right\} \text{X, Y hyperweak interactions}, \tag{10}$$

where we have indicated explicitly the strong interactions acting on the first three entries, the SU(2) weak interactions acting on the last two, and the hyperweak interactions mediated by X and Y bosons coupling together the first three and the last two indices. We see that the X and Y couple quarks to leptons, as well as quarks to antiquarks in the $\underline{10}$ representation not shown, and hence engender B violation and baryon decay. The simplest SU(5) model requires in addition two multiplets of Higgs fields, a $\underline{24} \phi$ with a vacuum expectation value $O(10^{15})$ GeV to break SU(5) to $SU(3) \times SU(2) \times U(1)$, and a $\bar{5}$ H with a vacuum expectation value $O(10^2)$ GeV to break $SU(2) \times U(1)$ to $U(1)_{em}$ and generate m_W, m_q and m_l.

Since there are a denumerable infinity of simple gauge groups, there are as many GUT models.

The next smallest one after SU(5) is SO(10) (Georgi 1975; Fritzsch & Minkowski 1975) which contains 45 gauge bosons, thereby offering baryons more ways to decay, three 16s of fermions, including a candidate for a right-handed neutrino, and at least three irreducible representations 16+45+10 of Higgses. Since this model introduces no fundamental new principles, we will concentrate on SU(5) as a bellwether GUT.

All GUTs predict charge quantization because they embed the U(1) of electromagnetism in a simple group, which means that charges are related by Clebsch–Gordan coefficients. The sum of the charges in every GUT representation must vanish, for example in the $\bar{5}$ of SU(5) we find from (10) that

$$\left.\begin{array}{l} Q_e = -1,\ 3Q_{\bar{d}} + Q_e = 0 \\ \Rightarrow Q_d = -\tfrac{1}{3} \Rightarrow Q_u = +\tfrac{2}{3} \Rightarrow \\ Q_p = 2Q_u + Q_d = +1, \end{array}\right\} \tag{11}$$

in accord with the experimental constraint (7).

GUTs also predict the 'fundamental' parameter $\sin^2 \theta_W$ (equation (2)) which is $\tfrac{3}{8}$ in the GUT symmetry limit $g_2 = g_1$, but gets renormalized in simple models as indicated in figure 1 to (Georgi et al. 1974; Marciano & Sirlin 1981; Llewellyn Smith et al. 1981)

$$\sin^2 \theta_W = 0.216 \pm 0.002, \tag{12}$$

to be compared (successfully) with the experimental value

$$\sin^2 \theta_W = 0.215 \pm 0.012, \tag{13}$$

as discussed by Llewellyn Smith (1983).

Another successful prediction of a 'fundamental' parameter which he did not mention is that of the b quark mass in terms of the τ lepton mass (Chanowitz et al. 1977). Generally, quark and lepton masses are related by Clebsch–Gordan coefficients in GUTs, and in minimal SU(5) we have $m_b = m_\tau$ in the GUT symmetry limit. This gets renormalized analogously to $\sin^2 \theta_W$, resulting in the physical prediction (Buras et al. 1978):

$$m_\tau = 1.78\,\text{GeV} \Rightarrow m_b \approx 5\,\text{GeV}, \tag{14}$$

if there are only six quarks in three generations (Nanopoulos & Ross 1979). In all fairness, it should be confessed that there are analogous predictions for m_s and m_d which are controversial and wrong respectively, but these may be modified without doing violence to the successful prediction (equation (14)).

The phenomenological successes (equations (7), (12) and (14)) constitute the only practical reasons so far for believing in grand unification, apart from its possible aesthetic appeal.

4. Baryon number violating interactions

After all this foreplay, let us get down to the nitty-gritty of GUTs, namely the prediction of baryon decay. The strength of the new interactions

$$G_X/2^{\frac{1}{2}} = g_X^2/8m_X^2, \tag{15}$$

yields a $\Delta B \neq 0$ amplitude $O(m_X^{-2})$ and hence a decay rate $O(m_X^{-4})$ and a lifetime

$$\tau_{p,n} = (m_X^4/m_B^5) \times O(1). \tag{16}$$

We will return shortly to the estimation of the $O(1)$ coefficient in (16), for the moment we just emphasize the great sensitivity to m_X which is estimated to be (Goldman & Ross 1980; Ellis *et al.* 1980*b*; Llewellyn Smith *et al.* 1981)

$$m_X = (1-2) \times 10^{15} \times \Lambda_{\overline{MS}} \tag{17}$$

in minimal SU(5), where as discussed by Llewellyn Smith (1983) a favoured range of $\Lambda_{\overline{MS}}$ is

$$100 \text{ MeV} < \Lambda_{\overline{MS}} < 200 \text{ MeV}, \tag{18}$$

though it could be as large as 400 MeV. Using (16)–(18) conventional estimates (Llewellyn Smith 1983) yield

$$\tau_{p,n} = 10^{29 \pm 2} \text{ years} \tag{19}$$

The favoured decay modes in minimal SU(5) are

$$\left. \begin{array}{l} p \to e^+\pi^0, \, e^+\omega, \, e^+\rho^0 \quad \text{and} \quad \mu^+K^0, \\ n \to e^+\pi^- \quad \text{and} \quad e^+\rho^-. \end{array} \right\} \tag{20}$$

Until recently, the predictions (19) and (20) looked quite healthy, there being several lower limits on the baryon lifetime of order 2×10^{30} years, but two reports (Krishnaswamy *et al.* 1982; Battistoni *et al.* 1982) of candidates for baryon decay, including one possible $p \to \mu^+K^0$. However, as discussed at this meeting by Goldhaber (1983), the I.M.B. collaboration (Bionta *et al.* 1983) has recently established that

$$\tau(p \to e^+\pi^0) > 10^{32} \text{ years}, \tag{21}$$

which is very embarrassing for conventional GUTs. They have one event compatible with $p \to \mu^+K^0$, but it could very well be a background neutrino interaction.

Does the result (21) rule out GUTs? There are many possible baryon decay modes that the I.M.B. collaboration has not yet searched for, and it clearly does not exclude models which do not predict the decay mode $p \to e^+\pi^0$, but the minimal conventional SU(5) described in §3 looks rather sick. We (Brodsky *et al.* 1983) have recently re-evaluated the baryon decay rate to be expected for a given value of m_X, i.e. the $O(1)$ coefficient in expression (16), in an attempt to answer the question at the head of this paragraph. We related the short distance baryon decay amplitude to knowledge about baryon wave functions at short distances gleaned from the proton magnetic form factor at large momentum transfers and from $J/\psi \to p\bar{p}$ decay. We found a much larger baryon decay rate than previous non-relativistic SU(6) and bag model calculations. This discrepancy may mean that the three-quark wavefunction overlap at short distance which controls the baryon decay rate in the chiral limit cannot be related easily to the one- and two-quark wavefunctions which may be known reliably from non-relativistic SU(6). Alternatively, it may mean that the proton form factor at $Q^2 = O(10)$ GeV^2 is not dominated by the short-distance three-quark baryon wave-function, which has accordingly been grossly over-estimated in the past (Isgur & Llewellyn Smith 1983, personal communication). This will undoubtedly become a controversy among practitioners of QCD. If we accept at face value our normalization of the baryon decay rate using form factor data, we infer from the limit (21) that

$$G_X < O(10^{-32}) \text{ GeV}^{-2} \Rightarrow m_X > 2 \times 10^{15} \text{ GeV}, \tag{22}$$

corresponding to $\Lambda_{\overline{MS}} > 1$ GeV according to the minimal SU(5) GUT relation (17). This makes the minimal SU(5) GUT look very sick indeed, even if it is 'not dead yet'.

Before abandoning this simple model, it is salutary to record (see table 1) how many 'fundamental' constants it contains. The gauge sector is certainly simpler than in the standard model: 1 gauge coupling g_5 instead of 3, and 1 non-perturbative vacuum parameter θ_5 instead of 2. There are still six quark masses, but now the three charged lepton masses are no longer independent of them. There are still three charged weak mixing angles, but now there are three CP-violating phases instead of one. The two new phases only appear in X and Y boson interactions and could play a role in cosmological baryosynthesis. There are now many more parameters needed to specify the boson masses, namely seven parameters in the minimal SU(5) Higgs potential. This model therefore has a total of 21 'fundamental' constants, which is not a significant improvement on the 20 of the standard model! Furthermore, the model predicts the wrong value of m_d, does not generate enough baryons in the early Universe, and probably predicts too short a baryon lifetime.

In fact, even before this latest experimental setback, the smart money had already moved out of stocks in minimal SU(5), as we see in the next section.

5. Attempts to understand the weak interaction scale

Since m_W at 80 GeV (Arnison *et al.* 1983a; Banner *et al.* 1983) is so much larger than the mass of any other known 'elementary' particle, it may seem at first sight strange to ask why m_W is so small. But m_W *is* very small on the scales of gravitation or of grand unification:

$$m_W/M_p = O(10^{-17}), \quad m_W/m_X \leqslant O(10^{-13})? \tag{23}$$

In spontaneously broken gauge theories m_W must be of the same order as the light Higgs boson m_H, which is awkward since the Higgs mass is notoriously unstable. We find

$$\delta m_H^2 = O(M_P^2), \tag{24}$$

from propagation through space–time foam at the Planck scale (Hawking *et al.* 1980), while couplings between the light Higgs H and the heavy Higgs ϕ in GUTs give

$$\delta m_H^2 = O(m_X^2), \tag{25}$$

from propagation through the GUT vacuum. Even if we set these to zero (how? why?), the Higgs mass is still destabilized by radiative corrections:

$$\delta m_H^2 = O(\alpha^n) \, (M_P^2 \text{ or } m_X^2). \tag{26}$$

This is the so called 'hierarchy problem' (Gildener & Weinberg 1976; Gildener 1976): we must understand what symmetry protects the light W and H from feedthroughs from the large mass scales. We now discuss two alternative strategies for such a 'solution' of the hierarchy problem. One may dissolve the offending diagrams by making H composite on a distance scale $x = O(1/1 \text{ TeV})$ and invoking dynamical symmetry breaking as in technicolour models. Alternatively one may cancel boson and fermion loops against each other as in supersymmetric theories.

(a) Technicolour

We postulate (Weinberg 1976, 1979; Susskind 1979) a complete new set of gauge interactions which become strong and confine unseen technifermions on a new distance scale $O(1/1 \text{ TeV})$. The previously elementary Higgs H is now replaced by a composite spinless techni-pion π_T which is a techni-fermion $\bar{F}F$ bound state, analogous to the conventional π which is a $(\bar{q}q)$ bound state:

$$H \to \pi_T = (\bar{F}F) \leftrightarrow \pi = (\bar{q}q). \tag{27}$$

The Higgs vacuum expectation value is replaced by a vacuum condensate as in QCD:

$$\langle 0| H |0\rangle \to \langle 0| \overline{F}F |0\rangle \leftrightarrow \langle 0| \bar{q}q |0\rangle, \tag{28}$$

which breaks weak gauge symmetry spontaneously:

$$m_W = g_2 \times O(\langle 0| \overline{F}F |0\rangle)^{\frac{1}{3}}, \tag{29}$$

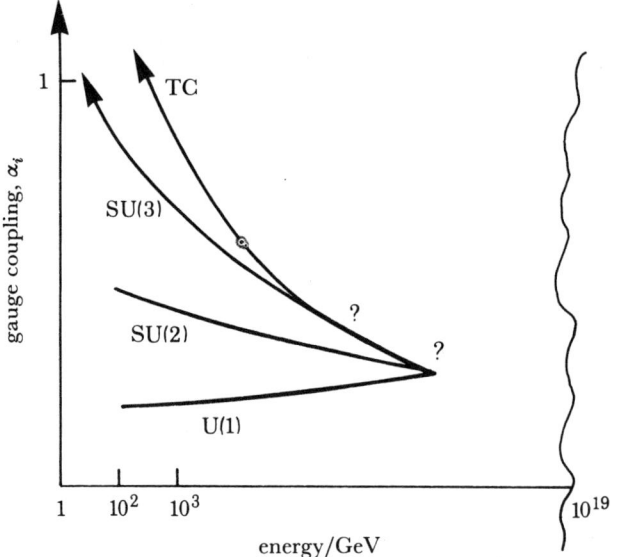

FIGURE 2. A sketch of technicolour (TC) interactions which get strong at a scale $O(1)$ TeV and may be unified with the other interactions at higher energies.

thanks to massless techni-pions being eaten by the $W^\pm$ and Z^0. One can imagine that the technicolour interaction is unified with the others at some high energy scale as seen in figure 2, in terms of which the scale of $\langle 0| \overline{F}F |0\rangle$ and hence m_W (equation (29)) is determined dynamically. Thus m_W and m_H would no longer be 'fundamental' constants.

This is a very economical scenario for generating m_W and m_Z, but to obtain non-zero masses for quarks and leptons we must add epicycles to this elegant theory. We need new extended technicolour interactions (Dimopoulos & Susskind 1979; Eichten & Lane 1980) mediated by new heavy gauge bosons E:

$$m_{q,1} = \langle 0| \overline{F}F |0\rangle / m_E^2. \tag{30}$$

These additional interactions cause problems, since there are related gauge boson exchanges which mediate flavour-changing interactions at levels far above experimental upper limits (Dimopoulos & Ellis 1981). Moreover, realistic theories contain many uneaten techni-pions that acquire calculable masses from conventional strong, weak and electromagnetic interactions, but not from extended technicolour interactions (Binétruy et al. 1981, 1982). None of these have been seen by experiment, which is disastrous in the case of the colourless charged techni-pions $P^\pm$ whose masses were calculated (Dimopoulos 1980; Chadha & Peskin 1981 a, b) to be

$$m_{P^\pm} \leq 15\,\text{GeV}. \tag{31}$$

This pair of disasters, coupled with the unattractive nature of the complicated extended technicolour interactions, has recently led to a general abandonment of technicolour.

(b) Supersymmetry

This is a new type of symmetry (Wess & Zumino 1974) in which fermions are connected to bosons:

$$Q|F\rangle = |B\rangle, \quad Q|B\rangle = |F\rangle, \tag{32}$$

by spinorial charges Q_α. As might be expected for fermionic objects, these charges Q_α obey an anticommutation algebra:

$$\{Q_\alpha^i, Q_j^{+\dot\alpha}\} = -2(\sigma^\mu)_\alpha^{\dot\alpha} \delta_j^i P_\mu, \tag{33}$$

where P_μ is the energy–momentum operator. In writing (33) we have sneaked in a new internal index $i = 1, 2, ..., N$ to label several different supersymmetry charges Q_α^i. The case $N = 1$ is simple supersymmetry, while $N > 1$ theories are said to possess extended supersymmetry. How large can N be? Renormalizable gauge theories are restricted to helicities $|\lambda| \leq 1$. Therefore at most four changes of spin $\frac{1}{2}$ by supersymmetry charges Q_α^i are permitted:

$$\lambda = +1 \xrightarrow{Q} +\tfrac{1}{2} \xrightarrow{Q} 0 \xrightarrow{Q} -\tfrac{1}{2} \xrightarrow{Q} -1; \tag{34}$$

and hence $N \leq 4$ for gauge theories. Supergravity theories with $|\lambda| \leq 2$ for the graviton are allowed to have $N \leq 8$ (van Nieuwenhuizen 1981). In most of what follows we will restrict ourselves to simple $N = 1$ supersymmetric theories, though some speculations about $N = 8$ supergravity will be advanced. The basic supermultiplets of $N = 1$ theories are the following

$$\text{gauge: } \lambda = \begin{pmatrix} 1 \\ \tfrac{1}{2} \end{pmatrix}; \quad \text{chiral: } \lambda = \begin{pmatrix} \tfrac{1}{2} \\ 0 \end{pmatrix}; \tag{35}$$

together with the graviton–gravitino $(2, \tfrac{3}{2})$ supermultiplet.

TABLE 2. SUPERSYMMETRIC PARTICLES

particle	spin	sparticle	spin
quark: q	$\tfrac{1}{2}$	squark: $\tilde{q}$	0
lepton: l	$\tfrac{1}{2}$	slepton: $\tilde{l}$	0
gluon: g	1	gluino: $\tilde{g}$	$\tfrac{1}{2}$
photon: γ	1	photino: $\tilde{\gamma}$	$\tfrac{1}{2}$
$W^\pm$	1	wino: $\tilde{W}$	$\tfrac{1}{2}$
Z^0	1	zino: $\tilde{Z}$	$\tfrac{1}{2}$
Higgs: H	0	shiggs: $\tilde{H}$	$\tfrac{1}{2}$

Unfortunately, no known particle can be the supersymmetric partner of any other, so we must at least double the number of known particles by the addition of unseen partners as seen in table 2. All the charged particles must have masses large enough to have avoided production and detection in e^+e^- collisions:

$$m_{\tilde{q}}, m_{\tilde{l}}, m_{\tilde{W}^\pm}, m_{\tilde{H}^\pm} \gtrsim O(17) \text{ GeV}. \tag{36}$$

The neutral ones could be rather lighter. For example, the best limit on the gluino mass from its absence in hadron–hadron collisions is (Bergsma et al. 1983)

$$m_{\tilde{g}} \gtrsim O(2) \text{ GeV}, \tag{37}$$

whereas particle physics offers no lower bound on the mass of the photino $\tilde{\gamma}$.

How heavy could these supersymmetric particles be? An answer is provided by attempts to stabilize the gauge hierarchy. The correspondence (equation (32)) between bosons and fermions

with identical couplings enforces systematic cancellations between loop diagrams which ensure that
$$\delta m_H^2 = O(\alpha) |m_B^2 - m_F^2|. \tag{38}$$

This is acceptably small, i.e. $O(m_W^2) = O(10^2\,\text{GeV})^2$ if
$$|m_B^2 - m_F^2| \lesssim O(1)\,\text{TeV}^2. \tag{39}$$

Thus the particles must be very light on the Planck scale, and potentially accessible to the next generation of particle accelerators. There are other phenomenological implications of supersymmetric theories beyond the existence of many new particles. For example, in minimal supersymmetric SU(5) GUTs (Dimopoulos & Georgi 1981; Sakai 1982) the prediction (12) for $\sin^2 \theta_W$ is modified (Ibáñez & Ross 1982; Einhorn & Jones 1982) to
$$\sin^2 \theta_W = 0.236 \pm 0.002, \tag{40}$$

which appears less successful. (Though the C.D.H.S. collaboration may soon announce a new, higher measurement of $\sin^2 \theta_W$.) A more dramatic modification is that of the GUT predictions (equations (19) and (20)) for baryon decay: in minimal supersymmetric GUTs (Dimopoulos *et al.* 1982; Ellis *et al.* 1982*d*)
$$p \to \bar{\nu} K^+, \quad \bar{\nu} K^{*+}; \quad n \to \bar{\nu} K^0, \quad \bar{\nu} K^{*0}, \tag{41}$$

with a less certain estimate of the total lifetime, thanks to our ignorance of the spectrum of supersymmetric particles. There is no longer a *prima facie* conflict with the negative results of the I.M.B. experiment.

So far, supersymmetric theories offer no clear explanation of the origin of m_W, but only allow a small value of m_W to be stabilized against radiative corrections. The first attempts (Dimopoulos & Georgi 1981; Sakai 1982) to construct supersymmetric theories neglected gravity and only used global supersymmetry. Spontaneously broken versions of these models came to grief, either because they had anomalies and/or did not break supersymmetry and/or had weak interactions becoming strong at energies much less than M_P (the so-called D theories) (Farrar & Weinberg 1983) or else required baroque spectra of unseen particles with unaesthetic symmetries imposed just to break supersymmetry (the so called F theories) (Ellis *et al.* 1982*a, b*). Therefore, recent model-building has focused on $N = 1$ supergravity theories with local supersymmetry. These theories embody the super-Higgs effect (Cremmer *et al.* 1979, 1983) which offers a new mechanism of spontaneous supersymmetry breaking not available in globally supersymmetric theories. In general they have a breaking scale
$$m_{\tilde{q}}, m_{\tilde{l}}, m_{\tilde{W}}, m_{\tilde{H}} = O(m_{\text{gravitino}}), \tag{42}$$

in the sparticle spectrum. If one neglects radiative corrections, the scale of weak gauge symmetry breaking is of the same magnitude
$$m_W = O(m_{\text{gravitino}}), \tag{43}$$

and this could well also be the case in theories where gauge symmetry breaking is induced by radiative corrections in the supergravity theory (Ellis *et al.* 1983*b*; Alvarez-Gaumé *et al.* 1983; Ibáñez & López 1983; Ellis *et al.* 1983*a*). However, it is also possible to construct models (Ellis *et al.* 1983*a*) with symmetry breaking by radiative corrections in which the weak gauge symmetry breaking scale is determined dynamically by dimensional transmutation, as described in more detail in the next section. In this case it is possible that
$$m_W \gg m_{\text{gravitino}}, \tag{44}$$

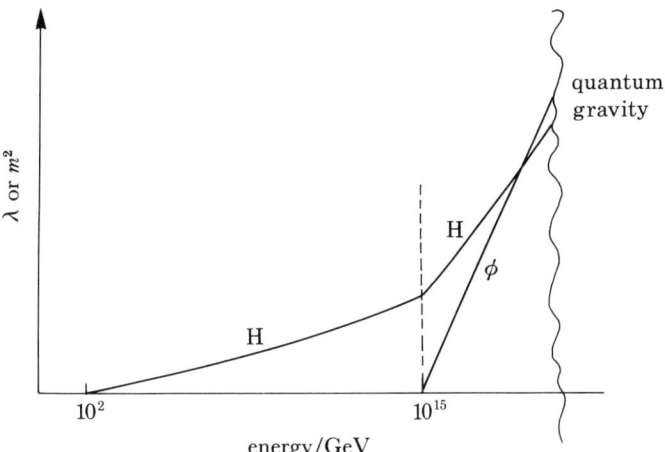

FIGURE 3. The coupling (mass) of a large Higgs representation ϕ evolves more rapidly than that of a small Higgs representation H. The grand unification symmetry may be broken when $\lambda_\phi(m_\phi)$ goes to zero, while weak SU(2) is broken when $\lambda_H(m_H)$ vanishes.

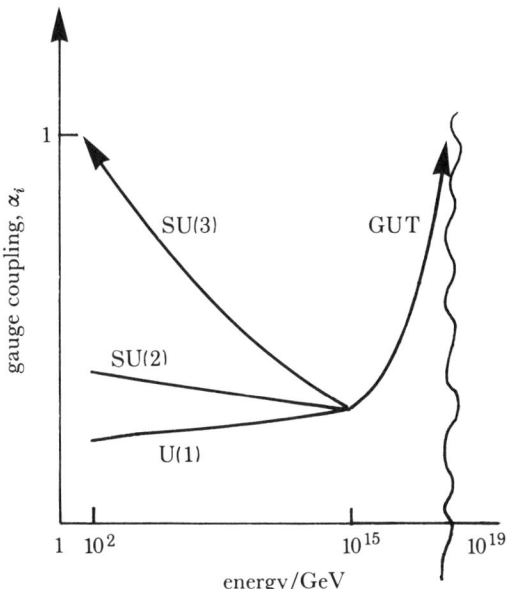

FIGURE 4. If there are enough particles with masses between 10^{15} and 10^{19} GeV, the strong coupling may decrease from $O(1)$ at M_P down to $O(\frac{1}{20} \text{ to } \frac{1}{40})$ at m_X.

though phenomenological considerations tell us that the supersymmetry breaking scale and hence $m_{\text{gravitino}}$ cannot be much less than 20 GeV. Models of weak gauge symmetry breaking by radiative corrections require the existence of at least one heavy fermion, and the most natural candidate would be the t quark. It would need to have a mass

$$m_t \gtrsim O(60) \text{ GeV}, \tag{45}$$

in these radiative scenarios.

It is an unfortunate feature of all these supergravity models that they require a light gravitino of mass $\lesssim O(m_W)$. There is as yet no clear idea how such a small mass parameter could emerge from a theory whose natural dynamical scale is $O(M_P)$. Thus supersymmetric theories have so far only lengthened, not shortened our list of 'fundamental' parameters.

[79]

6. ATTEMPTS TO UNDERSTAND THE GRAND UNIFICATION SCALE

In the simple models of grand unification discussed up till now, including supersymmetric ones, there is a grand unification scale $m_X \ll M_P$. It is possible to push m_X up to M_P only at the expense of introducing large numbers of additional low-mass fields (oases in the desert). Assuming that m_X is significantly less than M_P, it is interesting to speculate why m_X is so near to M_P compared with m_W on a logarithmic scale, and yet so far: $m_X/M_P = O(10^{-4})$? Possible scenarios for understanding the value of m_X are provided by the idea of dimensional transmutation alluded to earlier. In its original form due to Coleman & Weinberg (1973), the spontaneous breaking of weak gauge symmetry in a theory with zero Higgs mass at the tree level occurred at a mass scale where the renormalization group equations drove to zero a dimensionless parameter, namely a quartic Higgs self-coupling λ. If one imagines the initial value of this parameter being determined to be $O(\alpha)$ by dynamics at the Planck scale, then one finds that

$$m_W = M_P \exp(-O(1)/\alpha), \tag{46}$$

which is reminiscent of (8). In supersymmetric theories λ is unrenormalized, but its role can be usurped (Ellis *et al.* 1983*a*) by a combination m of the supersymmetry breaking Higgs masses in the theory, with the result (44). These ideas can be extended (Ellis *et al.* 1979, 1983*a*) to GUTs with two Higgs representations ϕ and H, the former large (24 of SU(5)?) and responsible for the initial GUT beaking at 10^{15} GeV, while the latter is smaller (5 of SU(5)?) and responsible for weak symmetry breaking at 10^2 GeV. One can imagine specifying a theory with $\lambda_{24} = O(\lambda_5)$ ($m_{24} = O(m_5)$) at the Planck scale. As one comes down to lower mass scales, $\lambda_{24}(m_{24})$ will change more rapidly because of the larger Casimir coefficients associated with larger group representations and calculations (Ellis *et al.* 1979, 1983*a*) indicate that it could easily vanish at $O(10^{-4}) M_P$ as required to fix m_X satisfactorily (see figure 3). Meanwhile, $\lambda_5(m_5)$ is non-zero and evolves even more slowly at lower mass-scales as indicated in figure 3, because the H representation is split in mass resulting in even smaller Casimir coefficients. Eventually $\lambda_5(m_5)$ will also vanish and thereby generate m_W, but because of the different group-theoretical coefficients it is very likely that

$$m_W/m_X \ll m_X/M_P = O(10^{-4}) \ll 1, \tag{47}$$

as desired.

It would also be nice to understand the gauge coupling α_X at the grand unification scale, which is about $\frac{1}{42}$ in minimal GUTs but about $\frac{1}{24}$ in supersymmetric GUTs. It is natural to suppose that $\alpha_X = O(1)$ at the Planck mass, which could facilitate full unification with gravity, possibly in an $N = 8$ supergravity theory as we shall speculate in a moment. If there are very many heavy particles with masses between m_X and M_P, their effect on the GUT renormalization group equations can be so large as to reverse the normal asymptotic freedom trend for α_X to increase with decreasing energy, and instead make α_X decrease as the energy scale decreases as shown in figure 4. If the contributions of these conjectured heavy particles in the renormalization group equations for α_X have about three times the magnitude of the gauge boson couplings driving asymptotic freedom, then α_X can (Ellis *et al.* 1982*c*) decrease from $O(1)$ at the Planck scale to $O(\frac{1}{20}$ to $\frac{1}{40})$ at the grand unification scale, as desired.

It is appropriate to speculate at the end of this talk about the possible eventual unification of all interactions at the Planck mass. The most natural candidate theory is supergravity, most probably the largest version with $N = 8$. We have already discussed the use of $N = 1$ supergravity, but this was as a phenomenological framework (Ellis *et al.* 1983*b*) for low energy physics at energy scales much less than M_P. If one takes a more fundamental approach and regards the $N = 8$ extended supergravity theory as an underlying theory of all elementary particle interactions (Ellis *et al.* 1980*a*; Ellis *et al.* 1980*c*), one encounters problems since all the known particles cannot be among the elementary states in the $N = 8$ graviton supermultiplet (Gell-Mann 1977). Perhaps some or all of the particles we know are not in fact 'elementary' but are actually composites of these $N = 8$ supergravity 'preons' (Cremmer & Julia 1979). While this is an attractive conjecture, our ignorance of supergravity dynamics does not yet allow us to put this idea on any sort of calculational basis, and there are even arguments that it may fail (Davis *et al.* 1983). If such a strategy could be made to work, it would offer the prospect of a theory with 0 (or perhaps 1) free parameter. This would at last be a significant reduction in the number of 'fundamental' constants.

References

Altarev, I. S. *et al.* 1981 *Physics Lett.* B **102**, 13.
Alvarez-Gaumé, L., Polchinski, J. & Wise, M. B. 1983 Harvard preprint HUTP-82/A063.
Arnison, G. *et al.* 1983*a* *Physics Lett.* B **122**, 103.
Arnison, G. *et al.* 1983*b* *Physics Lett.* B **126**, 398.
Banner, M. *et al.* 1983 *Physics Lett.* B **122**, 476.
Battistoni, G. *et al.* 1982 *Physics Lett.* B **118**, 461.
Bergsma, F. *et al.* 1983 *Physics Lett.* B **121**, 429.
Binétruy, P., Chadha, S. & Sikivie, P. 1981 *Physics Lett.* B **107**, 425.
Binétruy, P., Chadha, S. & Sikivie, P. 1982 *Nucl. Phys.* B **207**, 505.
Bionta, R. M. *et al.* 1983 *Phys. Rev. Lett.* **91**, 27.
Brodsky, S. J., Ellis, J., Hagelin, J. S. & Sachrajda, C. T. 1983 SLAC-PUB-3141.
Buras, A. J., Ellis, J., Gaillard, M. K. & Nanopoulos, D. V. 1978 *Nucl. Phys.* B **135**, 66.
Chadha, S. & Peskin, M. E. 1981*a* *Nucl. Phys.* B **185**, 61.
Chadha, S. & Peskin, M. E. 1981*b* *Nucl. Phys.* B **187**, 541.
Chanowitz, M. S., Ellis, J. & Gaillard, M. K. 1977 *Nucl. Phys.* B **128**, 506.
Coleman, S. & Weinberg, E. 1973 *Phys. Rev.* D **7**, 1888.
Cremmer, E., Ferrara, S., Girardello, L. & van Proeyen, A. 1983 *Nucl. Phys.* B **212**, 413.
Cremmer, E. & Julia, B. 1979 *Nucl. Phys.* B **159**, 141.
Cremmer, E., Julia, B., Scherk, J., Ferrara, S., Girardello, L. & van Nieuwenhuizen, P. 1979 *Nucl. Phys.* B **147**, 105.
Davis, A. C., Macfarlane, A. J. & van Holten, J. W. 1983 *Physics Lett.* B **125**, 151.
Dimopoulos, S. 1980 *Nucl. Phys.* B **168**, 69.
Dimopoulos, S. & Ellis, J. 1981 *Nucl. Phys.* B **182**, 505.
Dimopoulos, S. & Georgi, H. 1981 *Nucl. Phys.* B **193**, 150.
Dimopoulos, S., Raby, S. & Wilczek, F. 1982 *Physics Lett.* B **112**, 133.
Dimopoulos, S. & Susskind, L. 1979 *Nucl. Phys.* B **155**, 237.
Eichten, E. & Lane, K. D. 1980 *Physics Lett* B. **90**, 125.
Einhorn, M. B. & Jones, D. R. T. 1982 *Nucl. Phys.* B **196**, 475.
Ellis, J., Gaillard, M. K., Maiani, L. & Zumino, B. 1980*a* *Unification of the fundamental particle interactions* (ed. S. Ferrara, J. Ellis & P. van Nieuwenhuizen). New York: Plenum Press.
Ellis, J., Gaillard, M. K., Nanopoulos, D. V. & Rudaz, S. 1980*b* *Nucl. Phys.* B **176**, 61.
Ellis, J., Gaillard, M. K., Peterman, A. & Sachrajda, C. T. 1979 *Nucl. Phys.* B **164**, 253
Ellis, J., Gaillard, M. K. & Zumino, B. 1980*c* *Physics Lett.* B **94**, 343.
Ellis, J., Gaillard, M. K. & Zumino, B. 1982*c* *Acta phys. pol.* B **13**, 253.
Ellis, J., Hagelin, J. S., Nanopoulos, D. V. & Tamvakis, K. A. 1983*a* *Physics Lett.* B **125**, 275.
Ellis, J., Ibáñez, L. E. & Ross, G. G. 1982*a* *Physics Lett.* B **113**, 283.
Ellis, J., Ibáñez, L. E. & Ross, G. G. 1982*b* *Nucl. Phys.* B **221**, 29.
Ellis, J. & Nanopoulos, D. V. 1981 *Nature, Lond.* **292**, 436.
Ellis, J., Nanopoulos, D. V. & Rudaz, S. 1982*d* *Nucl. Phys.* B **202**, 43.
Ellis, J., Nanopoulos, D. V. & Tamvakis, K. A. 1983*b* *Phys. Lett.* B **121**, 123.
Farji, E. & Susskind, L. 1981 *Physics Rep.* C **74**, 277.
Farrar, G. & Weinberg, S. 1983 *Phys. Rev.* D **27**, 2732.
Fayet, P. & Ferrara, S. 1977 *Physic Rep.* C **32**, 249.

Fritzsch, H. & Minkowski, P. 1975 *Ann. Phys.* **93**, 193.
Gell-Mann, M. 1977 (Unpublished.)
Georgi, H. 1975 *Particles and Fields* 1974 (ed. C. E. Carlson). New York: A.I.P.
Georgi, H. & Glashow, S. L. 1974 *Phys. Rev. Lett.* **32**, 438.
Georgi, H., Quinn, H. & Weinberg, S. 1974 *Phys. Rev. Lett.* **33**, 451.
Gildener, E. 1976 *Phys. Rev.* D **14**, 1667.
Gildener, E. & Weinberg, S. 1976 *Phys. Rev.* D **13**, 3333.
Goldhaber, M. 1983 *Phil. Trans. R. Soc. Lond.* A **310**, 225.
Goldman, T. J. & Ross, D. 1980 *Nucl. Phys.* B **171**, 273.
Hawking, S., Page, D. N. & Pope, C. 1980 *Nucl. Phys.* B **170** (FS1), 283.
Ibáñez, L. E. & López, C. 1983 *Physics Lett.* B **126**, 54.
Ibáñez, L. E. & Ross, G. G. 1982 *Physics Lett.* B **105**, 439.
Isgur, N. & Wise, M. B. 1982 *Physics Lett.* B **117**, 179.
Kobayashi, M. & Maskawa, T. 1973 *Prog. theor. Phys.* **49**, 652.
Krishnaswamy, M. R. *et al.* 1982 *Physics Lett.* B **115**, 349.
Llewellyn Smith, C. H. 1983 *Phil. Trans. R. Soc. Lond.* A **310**, 253.
Llewellyn Smith, C. H., Ross, G. G. & Wheater, J. F. 1981 *Nucl. Phys.* B **177**, 263.
Marciano, W. J. & Sirlin, A. 1981 *Phys. Rev. Lett.* **46**, 163.
Nanopoulos, D. V. & Ross, D. A. 1979 *Nucl. Phys.* B **157**, 273.
Sakai, N. 1982 *Z. Phys.* C **11**, 153.
Susskind, L. 1979 *Phys. Rev.* D **20**, 2619.
van Nieuwenhuizen, P. 1981 *Physics Rep.* C **68**, 189.
Weinberg, S. 1976 *Phys. Rev.* D **13**, 974.
Weinberg, S. 1979 *Phys. Rev.* D **19**, 1277.
Weinberg, S. 1983 *Phil. Trans. R. Soc. Lond.* A **310**, 249.
Wess, J. & Zumino, B. 1974 *Nucl. Phys.* B **70**, 39.

Phase transitions in the early Universe and their consequences

By T. W. B. KIBBLE, F.R.S.

Blackett Laboratory, Imperial College, Prince Consort Road, London SW7 2BZ, U.K.

The reasons for believing that a number of phase transitions occurred in the early Universe are reviewed, and their implications discussed. In particular, the current status of the explanation for the observed values of some constants in terms of the 'inflationary universe' is examined.

INTRODUCTION

The observed cosmic redshift and microwave background provide good evidence that at earlier times the Universe was denser and hotter than it now is. Beyond a redshift Z of 1300, corresponding to a temperature of 4000 K, it was filled with a dense plasma of ionized hydrogen, more or less in thermal equilibrium though expanding adiabatically.

At yet earlier times, $t \lesssim 1$ s, when thermal energies exceeded 1 MeV, elementary particle processes dominated. The relevant equation of state then is more speculative. However, if our present ideas about fundamental particle interactions are on the right lines, then it is in fact a reasonable approximation over much of the earlier period to treat the matter in the Universe as a weakly-interacting relativistic gas. In particular, if the strong interactions are indeed describable by quantum chromodynamics (QCD) then asymptotic freedom suggests that at sufficiently high temperature all particle interactions are effectively weak.

PHASE TRANSITIONS

We may additionally expect the smooth adiabatic expansion to be interrupted by a number of phase transitions. (For a discussion and earlier references see Kibble (1980) or Kibble (1982).) The currently accepted theory of electromagnetic and weak interactions – the Weinberg–Salam model – predicts a phase transition at a temperature of order 100 GeV at which the $SU(2) \times U(1)$ symmetry breaks spontaneously to the $U(1)$ of electromagnetic gauge invariance. Moreover in QCD there is almost certainly a phase transition in the vicinity of 100 MeV at which confinement sets in. Above this temperature we have a gas of unconfined quarks and gluons; below, a gas of hadrons.

We can be fairly confident about these two transitions. Both may be accompanied by interesting phenomena of various kinds, but neither is likely to introduce any really dramatic disturbance to the smooth evolution of the Universe. What happens at yet higher temperatures is far more problematic, because we really know very little about particle interactions beyond 1 TeV. However there are good reasons for supposing that other phase transitions occurred.

Since both strong and electroweak interactions are well described by gauge theories, the idea of grand unification (Georgi & Glashow 1974) seems very appealing. In a grand unified theory (GUT) there is a phase transition (or perhaps several) at a temperature T_c of, typically, about 10^{15} GeV. Above this, the equilibrium state is fully symmetric under a group such as $SU(5)$ or $SO(10)$, while at lower temperatures the symmetry breaks to $SU(3)_{\text{colour}} \times [SU(2) \times U(1)]_{\text{WS}}$.

Unlike the later transitions, this one is probably strongly first-order (Guth & Tye 1980; Lazarides & Shafi 1980; Cook & Mahanthappa 1981; Guth & Weinberg 1981; Einhorn & Sato 1981; Billoire & Tamvakis 1981). We may expect the Universe to supercool in its (metastable) symmetric phase. The equilibrium states correspond to minima of the effective potential or free-energy density $V(\phi)$, where ϕ is some order parameter or Higgs field. In this case, V has two minima separated by a barrier. Below T_c the minimum at $\phi = 0$ is no longer the absolute minimum, but the Universe remains there until it can tunnel through the barrier. During this supercooling the energy density comes to be dominated by the vacuum energy, $V(\phi)$, which is of order T_c^4. (The zero is chosen to be at the absolute minimum of V.) This leads to an exponential expansion,

$$R(t) \propto e^{Ht}, \quad \text{where } H \text{ is of order } T_c^2/M_P. \quad (M_P \text{ is the Planck mass}).$$

This is the *inflationary universe* (Guth 1981).

If the transition is terminated by nucleation of bubbles of the new phase, one finds essentially all the energy concentrated in the bubble walls and hence an impossibly inhomogeneous distribution of matter (Einhorn *et al.* 1980; Guth & Weinberg 1981; Hawking *et al.* 1982). This problem has been at least partly overcome in the *new inflationary universe* (Linde 1982a; Albrecht & Steinhardt 1982; Hawking & Moss 1982). In this model, $V(\phi)$ is extremely flat near $\phi = 0$, with no ϕ^2 term. Consequently, even after the transition, ϕ will take a long time to 'roll down' the curve of V towards its absolute minimum. This has two effects. First, exponential expansion will continue for a long time after the transition; thus the bubble size will expand by an enormous factor, so that our present Universe is entirely contained within a single bubble. Second, the energy will not be deposited in the bubble wall but will spread much more uniformly, suggesting that we might avoid the problem of excessive inhomogeneity.

As we shall see, there are still problems with this scenario. In particular, the condition that the potential be very flat requires an extremely precise fine tuning of parameters. However it has been suggested by several authors (Ellis *et al.* 1982; Albrecht *et al.* 1982; Vayonakis 1983) that inflation may occur naturally in a supersymmetric model. There are also difficulties, to be discussed later, concerning the magnitude of the fluctuations in the new inflationary universe.

Another problem that has not been fully resolved concerns the gravitational effects on the phase transition. During the period of exponential expansion the Universe is effectively in a de Sitter space, and once the temperature falls to the associated Hawking temperature T_H, of order H, the curvature corrections to the effective potential become important (Abbott 1981; Hut & Klinkhamer 1981; see also Shore 1980; Pollock & Calvani 1982; Brandenberger & Khan 1982).

DENSITY OF THE EARLY UNIVERSE

Let us concentrate on the Universe just before the electroweak phase transition, when it is say 1 ps old and in thermal equilibrium at a temperature of about 1 TeV. From that point on, we can follow its evolution reasonably well. What we have to understand, therefore, is the initial state at that time.

Let us consider the various parameters that will need to be specified. From the observed isotropy of the microwave background, which is isotropic to better than one part in 10^3, we may conclude that the Universe must have been to a good approximation homogeneous and isotropic. Thus it may be represented at least approximately by one of the Robertson–Walker metrics

$$ds^2 = dt^2 - R^2(t)\, d\sigma^2, \tag{1}$$

where $d\sigma^2$ is a 3-metric of uniform curvature K.

The rate of expansion $\dot{R}$, or the Hubble parameter $H = \dot{R}/R$, is related to the density ρ by Einstein's equation

$$H^2 = (\dot{R}/R)^2 = (8\pi/3M_{\rm P}^2)\,\rho - K/R^2 - \Lambda, \tag{2}$$

where Λ is the cosmological constant and $M_{\rm P} = G^{-\frac{1}{2}}$ is the Planck mass. Observationally, Λ is very small, consistent with zero:

$$\Lambda/M_{\rm P}^2 \lesssim 10^{-122}. \tag{3}$$

It is for this reason that the absolute minimum of the effective potential V is taken to be at zero. This may find a natural explanation within the context of a supergravity model, where the energy is positive definite (Deser & Teitelboim 1977), and radiative corrections to Λ may be expected to cancel. However it is very hard to see how such a result would survive supersymmetry breaking. One possible explanation will be discussed in the following paper (Hawking 1983).

If $\Lambda = 0$, the critical density $\rho_{\rm c}$ that separates closed and open universes is given by

$$\rho_{\rm c} = 3M_{\rm P}^2 H^2/8\pi.$$

If $\rho \leqslant \rho_{\rm c}$ then $K \leqslant 0$; the Universe is open and will continue expanding for ever. If $\rho > \rho_{\rm c}$, then $K > 0$. In that event the Universe is closed; it will reach a maximum size and then start to recontract.

As the Universe evolves, ρ varies initially like R^{-4}, while it is dominated by relativistic particles, and eventually like R^{-3} when rest-mass is the largest contribution. On the other hand, from Einstein's equation $\rho - \rho_{\rm c}$ varies much more slowly, like R^{-2}. In the Universe today, ρ is within about an order of magnitude of $\rho_{\rm c}$. To achieve this, it must have been much closer in the early Universe. In fact at the time we have chosen

$$|\rho - \rho_{\rm c}|/\rho \lesssim 10^{-26}; \tag{4}$$

this is another parameter we have to understand.

One of the major successes of the new inflationary universe is that it provides a natural explanation for this very small number. For, during the period of exponential expansion ρ is essentially constant (equal to $V(0)$) while as before $\rho - \rho_{\rm c} \propto R^{-2}$. In this situation, $\rho_{\rm c}$ is driven towards, rather than away from, ρ. It is easy to arrange that during inflation R increases by 10^{28} or more, yielding a very adequate reduction in $\rho - \rho_{\rm c}$.

Baryon and lepton numbers

To specify a thermal equilibrium state we need not only the temperature or density but also the values of any absolutely conserved quantum numbers. So far as physics below 1 TeV is concerned, the only such quantities we known of are electric charge Q, baryon number B and the various lepton numbers $L_{\rm e}$, L_{μ}, L_{τ}, though of course it is conceivable that there might be further families of leptons.

A homogeneous state of nonzero electric charge is impossible by Gauss's law, and we may therefore assume that on average Q is zero. On the other hand the net baryon number is not zero, certainly in our galaxy, and most likely in the Universe as a whole. It is most easily characterized by the baryon-to-photon ratio which is approximately constant during adiabatic expansion, or better still the baryon-to-entropy ratio n_b/s. Here n_b is the baryon-number density, and s is the entropy density, given to a good approximation by

$$s = (2\pi^2/45)\, N_* \, T^3,$$

where N_* is the effective number of species of massless particles (i.e. helicity states of bosons, plus those of fermions times $\frac{7}{8}$). This ratio has the advantage that it remains constant even during the various annihilation episodes when particle pairs annihilate and enhance the photon number, provided the processes are reversible.

From observations of matter in the Universe today, we find that

$$n_b/s \approx 10^{-11 \pm 1}. \tag{5}$$

This again is a parameter that needs explanation.

The lepton numbers are far less accurately known. It is not even wholly inconceivable that the ratio n_l/s could be of order 1, corresponding to a degenerate sea of neutrinos of some type (Langacker *et al.* 1982). The three corresponding lepton-to-entropy ratios complete the list of equilibrium parameters.

A period of inflation naturally leads to essentially zero values of all conserved quantities. Whatever their pre-existing values, they are diluted by an enormous factor.

The only parameter that is definitely non-zero is the baryon-to-entropy ratio. It was one of the first great successes of the GUT idea that it provided a natural explanation for this non-zero value.

In the high-temperature symmetric phase of GUTS, baryon-number-violating processes occur quite freely. This is because at a fundamental level the quarks and leptons belong to the same multiplet. Transitions between them (which induce for example proton decay) are mediated by the exchange of superheavy X bosons. Since m_X is of order 10^{15} GeV, the transition rates are heavily suppressed except at very high temperatures.

We may imagine therefore that at very early times the Universe was in this high-temperature symmetric phase and that the observed non-zero mean baryon number was generated by irreversible CP-violating processes following the phase transition. These could be decay of X bosons or of the associated Higgs bosons, but there are other possibilities too: for example, the decay of Higgs-field fluctuations produced at a first-order transition (Hawking & Moss 1983; see also, Abbott *et al.* 1982; Dolgov & Linde 1982), the annihilation of monopoles or the decay of string loops (Bhattacharjee *et al.* 1982).

Although this explanation for a non-zero baryon number was, and indeed still is, one of the great successes of grand unification, it is important to recognize that it is a very partial success. It provides a qualitative understanding, but at present it cannot provide a quantitative estimate. All it can do is to relate one small parameter, the baryon-to-entropy ratio, to another, the fundamental CP-violation parameter ϵ. At present we have no means of calculating this parameter *a priori*.

Moreover many grand unified theories do not fully satisfy the so-called gauge principle, in that there remain some exact conservation laws corresponding to global invariances, for example a fermion number equivalent to conservation of $B-L$, or indeed separate fermion numbers for different generations. It is then an *assumption* that the mean value of this quantum number is zero in the intial symmetric phase.

Once again of course the new inflationary universe provides a natural explanation for the zero value of any such quantum number. It therefore does make a definite prediction that the lepton-to-entropy ratio should be equal to the baryon-to-entropy ratio. This is a distinctly non-trivial prediction that should in principle be testable. Moreover there are other predictions concerned with fluctuations to which I shall return.

Defects in the initial state

So far we have discussed only equilibrium parameters, but we know that the early Universe cannot have been precisely in a homogeneous equilibrium state. The statistical fluctuations in such a state would be far too small to yield the distribution of matter that we see today.

To complete the description of the state of the Universe at 1 ps we have to specify what inhomogeneities it contains and how it departs from the simple Robertson–Walker form. The inhomogeneities include both spatial fluctuations in the equilibrium parameters (and the metric) and also topological defects of various sorts.

Let us consider first the defects. Depending on the topology of the gauge group and its unbroken subgroup, a phase transition may generate defects of various spatial dimensions: monopoles, strings or domain walls. Monopoles appear naturally in almost all GUTs, while strings and domain walls appear in only certain models (Kibble 1976). In the case of multiple phase transitions, there are also composite structures, strings terminated by monopoles or walls bounded by strings (Bais 1981; Kibble *et al.* 1982 *a, b*). However these objects usually disappear quite quickly, and for simplicity I shall ignore them. I shall confine my discussion to topologically *stable* defects.

Domain walls within our presently visible Universe can be ruled out because their gravitational effects would induce enormous anisotropy (Zel'dovich *et al.* 1974). Until recently this argument has been used to exclude theories with stable domain walls corresponding to the breaking of an exact discrete symmetry. However the inflationary universe may rehabilitate such models, because any domain walls would probably be so far distant as to be undetectable.

Monopoles like baryons are best characterized by the monopole-to-entropy ratio n_m/s. The requirement that monopoles now contribute to the mean density of the universe no more than about the critical density ρ_c implies that at present $n_m/s \lesssim 10^{-24}$. Since very little annihilation can have occurred after our chosen initial time of 1 ps, this limit must apply then too (Preskill 1979; Zel'dovich & Khlopov 1978; Goldman *et al.* 1981). Another even stronger limit, $n_m/s \lesssim 10^{-25}$, is based on the survival of the galactic magnetic field (Turner *et al.* 1982, Lazarides *et al.* 1981; Arons & Blandford 1983). The new inflationary universe naturally predicts for monopoles too an essentially zero value, thus solving the 'cosmic monopole problem' (Guth 1981). Indeed if the reported detection of a monopole (Cabrera 1982) is confirmed, we should have to suppose that monopoles, like baryons, are made after the main period of inflation, for example at a subsequent monopole-generating phase transition. At first sight one might suppose that it would then again be very difficult to avoid an excess of monopoles. However, Moss (1983) has recently shown that this may not be so, in the context of an SU(5) model with a two-stage transition via SU(4) × U(1) to SU(3) × SU(2) × U(1).

There are other suggestions for reducing, the monopole density, for example the model of Langacker & Pi (1980) involving an intermediate phase where the monopoles disappear. But E. Weinberg (1983) has recently shown using very general causality arguments that all such models are perilously close to the observational limits.

Strings too would be reduced to essentially zero density by inflation. If they are to have any relevance to observational cosmology, they must be produced at a subsequent transition. However it is not at all difficult to meet the observational limits in this case.

As before, we may impose the requirement that their total contribution to the mean density of the Universe can at no time much exceed the critical density. Indeed, in recent times (i.e. since decoupling) it must have been some orders of magnitude less to avoid introducing unacceptably large anisotropy into the microwave background via gravitational effects.

The strings may be characterized at any time by the mean length of string per unit volume l. For a random (Brownian) configuration with persistence length ξ, one has $l \approx \xi^{-2}$. If μ is the mass per unit length, or tension, we require

$$\mu l \ll \rho. \tag{6}$$

It should be noted that if the configuration of strings simply expands conformally with the expansion of the Universe then

$$l \propto R^{-2},$$

where as ρ scales like R^{-3} or R^{-4}. Thus strings would soon come to dominate unless there were some mechanism to reduce l and transfer energy from strings to other matter or radiation.

However there are such mechanisms. In the very early stages, strings are heavily damped by interaction with surrounding matter, but this condition does not persist for long (Kibble 1976, 1982). Thereafter the most important mechanism appears to involve the formation of closed loops. When strings intersect they may exchange partners. Sometimes this process yields closed loops which then oscillate, gradually losing energy by gravitational (or perhaps other) radiation until they disappear (Vilenkin 1981; Kibble & Turok 1982; Bhattacharjee *et al.* 1982). It is not easy to estimate the rate of production of closed loops, but simple scaling arguments suggest that it should be sufficient to ensure that (after an initial period) the typical length scale ξ of strings remains a constant fraction of the horizon distance. This means that

$$l \approx \xi^{-2} \approx t^{-2}, \tag{7}$$

so that the ratio $\mu l/\rho$ is in fact approximately constant, and compatible with (6). It is possible that the small loops that survive for long periods may be of relevance to the problem of galaxy formation (Vilenkin 1981; Kibble & Turok 1982; Turok 1983).

Fluctuations

Let us now turn to the other type of inhomogeneity: spatial fluctuations in the equilibrium parameters.

It is customary (Peebles 1980; see also Bardeen 1980) to classify the small perturbations on the Friedman–Robertson–Walker universes in three classes: transverse gravitational waves, adiabatic perturbations (i.e. spatial fluctuations in temperature and density with constant baryon-to-entropy ratio) and isothermal fluctuations (spatial fluctuations on the baryon-to-entropy ratio at constant temperature). In fact there might be several sorts of isothermal fluctuations, because in principle there might be independent fluctuations in the three lepton-to-entropy ratios.

Gravitational waves are simply another type of radiation which, because they decouple rather early, seems not to be of great cosmological significance. I shall not consider them further.

One of the important predictions made by the standard GUT scenario for baryon-number generation is that there should be essentially no isothermal perturbations. This is because the baryon-number-generating mechanism depends only on the temperature and so will always yield the same baryon-number density at a given temperature. It would only be possible to escape this conclusion if we found a way to make the maximum reheating temperature vary from place to place, or, as Stephen Hawking pointed out to me, if the CP-violating parameter were variable.

The new inflationary universe also makes very definite predictions about the spectrum of adiabatic perturbations. It is convenient to characterize the magnitude of the perturbation by the value of the density contrast $\delta\rho/\rho$ at the moment when it comes within the horizon, $(\delta\rho/\rho)_h$, say. It has been shown by several different groups that the new inflationary universe generates a spectrum of perturbations for which $(\delta\rho/\rho)_h$ is almost constant, independent of wavelength (Hawking 1982; Albrecht & Steinhardt 1982; Starobinsky 1982; Vilenkin & Ford 1982; Linde 1982b; Guth & Pi 1982). This is very encouraging, because this is exactly the spectrum required by the Zel'dovich 'pancake' theory of galaxy formation (Zel'dovich 1970; Harrison 1970).

Unfortunately, though the shape is right the magnitude is substantially too large. However, several authors have suggested that this problem may be avoided in a supersymmetric inflationary model.

Albrecht et al. (1982, see also Banks & Kaplunovsky 1982; Pi 1982) proposed a model incorporating the O'Raiffeartaigh type of supersymmetry breaking (O'Raiffeartaigh 1975), as in the reverse-hierarchy mechanism (Witten 1981) or the geometric hierarchy (Dimopoulos & Raby 1983). This naturally yields a very slowly-varying potential, and because the inflation occurs away from $\phi = 0$ the fluctuations may be reduced in magnitude. However, for this type of symmetry breaking the minimum of the (positive-definite) tree-level potential occurs at a non-zero value, and it may therefore be difficult to reconcile with a naturally vanishing cosmological constant. It is also difficult to ensure adequate reheating to make baryons.

A rather different supersymmetric inflationary model due to Ellis et al. (1982) envisages a transition temperature very close to the Planck mass (see also Nanopoulos et al. 1983). This model is expected to yield acceptably small perturbations (Ellis et al. 1983), but on the other hand fails to solve the monopole problem.

Conclusions and discussion

Let us recall the various parameters that are required to fix the initial state at 1 ps and see how many of them can be understood or predicted by the model of phase transitions, and particarly the new inflationary universe.

The near-vanishing of the cosmological constant is still unexplained, though it may perhaps be easier to explain in the context of a supergravity model (Hawking 1983).

The inflationary universe naturally explains the small value of $(\rho-\rho_c)/\rho$, and indeed makes a definite prediction that within observational limits ρ should be exactly equal to ρ_c. It also predicts near-zero values for the densities of defects of all kinds that are produced at or before the inflationary transition, though it leaves open the possibility that strings (or perhaps monopoles) might be produced at a later transition.

There is a qualitative understanding of the baryon-to-entropy ratio, and a clear prediction that the lepton-to-baryon ratio should be unity.

The new inflationary universe naturally explains the overall homogeneity of the Universe, since all we can now observe has evolved from one single tiny bubble. So far as fluctuations are concerned, the predicted spectrum of adiabatic fluctuations has the right shape. The best hope of producing the right magnitude too seems to lie with supersymmetric models. In general one would not expect any isothermal fluctuations.

One thing I have not discussed is the state of the Universe *before* the inflationary phase transition. In fact we know very little about it. It is characteristic of inflation that it erases almost all information about the pre-existing state. The Universe may have been extremely homogeneous or extremely inhomogeneous, but since our observations are confined to what was the interior of a single tiny bubble, we have no means of knowing. It might also be that the transition occurred not by formation of a bubble but instead that the Universe as a whole made a transition to the new phase (Hawking & Moss 1982). Even more exotically, it has been suggested (Vilenkin 1982) that what existed before the transition was nothing at all, that our Universe was created literally from nothing! Though one can indeed perform a quantum-mechanical tunnelling-probability calculation for this conjectured process it is very hard to know how to interpret it!

In any event we cannot expect to go much further back in time without encountering quantum gravity effects (at around the Planck mass), and since there is as yet no wholly satisfactory theory of quantum gravity this is at present impossible.

What is remarkable, however, is the success of the new inflationary universe in explaining so many of the parameters that describe our Universe. This must give good ground for confidence that we are least on the right lines.

References

Abbott, L. F. 1981 *Nucl. Phys.* B **185**, 233.
Abbott, L. F., Farhi, E. & Wise, M. B. 1982 *Physics Lett.* B **117**, 329.
Albrecht, A., Dimopoulos, S., Fischler, W., Kolb, E., Raby, S. & Steinhardt, P. J. 1982 *Proceedings of the 3rd Marcel Grossmann Conference.* (In preparation.)
Albrecht, A. & Steinhardt, P. J. 1982 *Phys. Rev. Lett.* **48**, 1220.
Arons, J. & Blandford, R. D. 1983 *Phys. Rev. Lett.* **50**, 544.
Bais, F. A. 1981 *Physics Lett.* B **98**, 437.
Banks, T. & Kaplunovsky, V. 1982 *Nucl. Phys.* B **206**, 45.
Bardeen, J. M. 1980 *Phys. Rev.* D **22**, 1882.
Bhattacharjee, P., Kibble, T. W. B. & Turok, N. 1982 *Physics Lett.* B **119**, 95.
Billoire, A. & Tamvakis, K. 1981 CERN preprint TH.3019-CERN, unpublished.
Brandenberger, R. & Kahn, R. 1982 *Physics Lett.* B **119**, 75.
Cabrera, B. 1982 *Phys. Rev. Lett.* **48**, 1378.
Cook, G. P. & Mahanthappa, K. T. 1981 *Phys. Rev.* D **23**, 1321.
Deser, S. & Teitelboim, C. 1977 *Phys. Rev. Lett.* **39**, 249.
Dimopoulos, S. & Raby, S. 1983 *Nucl. Phys.* B (Submitted.)
Dolgov, A. D. & Linde, A. D. 1982 *Physics Lett.* B **116**, 329.
Einhorn, M. B. & Sato, K. 1981 *Nucl. Phys.* B **180**, 385.
Einhorn, M. B., Stein, D. L. & Toussaint, D. 1980 *Phys. Rev.* D **21**, 3295.
Ellis, J., Nanopoulos, D. V., Olive, K. A. & Tamvakis, K. 1982 *Physics Lett.* B **118**, 335.
Ellis, J., Nanopoulos, D. V., Olive, K. A. & Tamvakis, K. 1983 *Physics Lett.* B **120**, 331.
Georgi, H. & Glashow, S. 1974 *Phys. Rev. Lett.* **32**, 438.
Goldman, T., Kolb, E. & Toussaint, D. 1981 *Phys. Rev.* D **23**, 867.
Guth, A. H. 1981 *Phys. Rev.* D **23**, 347.
Guth, A. H. & Pi, S.-Y. 1982 *Phys. Rev. Lett.* **49**, 1110.
Guth, A. H. & Tye, S. H. 1980 *Phys. Rev. Lett.* **44**, 631.
Guth, A. H. & Weinberg, E. J. 1981 *Phys. Rev.* D **23**, 876.
Harrison, E. R. 1970 *Phys. Rev.* D **1**, 2726.
Hawking, S. W. 1982 *Physics Lett.* B **115**, 295.
Hawking, S. W. 1983 *Phil. Trans. R. Soc. Lond.* A **310**, 303.
Hawking, S. W. & Moss, I. G. 1982 *Physics Lett.* B **110**, 35.
Hawking, S. W. & Moss, I. G. 1983 Fluctuations in the inflationary universe (Unpublished preprint.)

Hawking, S. W., Moss, I. G. & Stewart, J. M. 1982 *Phys. Rev.* D **26**, 2681.
Hut, P. & Klinkhamer, F. R. 1981 *Physics Lett.* B **104**, 439.
Kibble, T. W. B. 1976 *J. Phys.* A **9**, 1387.
Kibble, T. W. B. 1980 *Physics Rep.* **67**, 183.
Kibble, T. W. B. 1982 *Acta Phys. pol.* B **13**, 723.
Kibble, T. W. B., Lazarides, G. & Shafi, Q. 1982a *Physics Lett.* B **113**, 237.
Kibble, T. W. B., Lazarides, G. & Shafi, Q. 1982b *Phys. Rev.* D **26**, 435.
Kibble, T. W. B. & Turok, N. 1982 *Physics Lett.* B. **116**, 141.
Langacker, P. & Pi, S.-Y. 1980 *Phys. Rev. Lett.* **45**, 1.
Langacker, P., Segrè, G. & Soni, S. 1982 *Phys. Rev.* D **26**, 3425.
Lazarides, G. & Shafi, Q. 1980 Unpublished.
Lazarides, G., Shafi, Q. & Walsh, T. F. 1981 *Nucl. Phys.* B **195**, 157.
Linde, A. D. 1982a *Physics Lett.* B **108**, 389.
Linde, A. D. 1982b *Physics Lett.* B **116**, 335.
Moss, I. G. 1983 *Monopoles in the inflationary universe*. Newcastle University preprint (unpublished).
Nanopoulos, D. V., Olive, K. A., Srednicki, M. & Tamvakis, 1983 *Physics Lett.* **123**, 41.
O'Raiffeartaigh, L. 1975 *Nucl. Phys.* B **96**, 331.
Peebles, P. J. A. 1980 *The large-scale structure of space–time*. Princeton University Press.
Pi, S.-Y. 1982 *Physics Lett.* B **112**, 441.
Pollock, M. D. & Calvani, M. 1982 *Physics Lett.* B **117**, 392.
Preskill, J. P. 1979 *Phys. Rev. Lett.* **43**, 1365.
Shore, G. M. 1980 *Ann. Phys.* **128**, 376.
Starobinsky, A. A. 1982 *Physics Lett.* B **117**, 175.
Turner, M. S., Parker, E. N. & Bogdan, T. J. 1982 *Phys. Rev.* D **26**, 1296.
Turok, N. 1983 *Physics Lett.* B **123**, 387.
Vayonnakis, C. E. 1983 *Physics Lett.* B **123**, 396.
Vilenkin, A. 1981 *Phys. Rev. Lett.* **46**, 1169, 1496(E).
Vilenkin, A. 1982 *Physics Lett.* B **117**, 25.
Vilenkin, A. & Ford, L. H. 1982 *Phys. Rev.* D **26**, 1231.
Weinberg, E. J. 1983 Columbia University preprint CU-TP-262 (unpublished).
Witten, E. 1981 *Physics Lett.* B **105**, 267.
Zel'dovich, Ya. B. 1970 *Astron. Astrophys.* **5**, 84.
Zel'dovich, Ya. B. & Khlopov, M. Y. 1978 *Physics Lett.* B **79**, 239.
Zel'dovich, Ya. B., Kobzavev, I. Ya. & Okun, L. B. 1974 *Zh. èksp. teor. Fiz.* **67**, 3 (*Soviet Phys. JETP* **40**, 1).

Discussion

S. KASDAN (*Imperial College, Prince Consort Road, London SW 7 2BZ, U.K.*). There is now a growing body of literature – by a growing body I mean more than two papers – that establishes that the effective potential, $V(\phi)$, does not exist as a calculable quantity for a range of ϕ around the origin when symmetry breaking takes place and ϕ acquires a non-zero vacuum expectation value. Results by Callaway & Maloof (1983) and Haymaker & Perez-Mercader (1983) show that possible broken symmetry phases cannot be compared by calculating $V(\phi)$ because $V(\phi)$ is actually undefined in the region of interest. Therefore, one cannot determine, say, the true vacuum by an apparently calculable lowest minimum of $V(\phi)$.

T. W. B. KIBBLE. I accept what Dr Kasdan says, but I regard that as an essentially technical problem. I would be surprised if it affected the physical results.

S. KASDAN. It is more than a technical problem. One is attempting to determine the existence and direction of possible phase transitions by looking at the energy differences between minima of $V(\phi)$ and these results (Callaway & Maloof 1983; Haymaker & Perez-Mercader 1983) imply that this is really a meaningless or undecidable question. It is interesting that this problem was actually first studied in perturbation theory by Coleman & Weinberg (1973) when they developed effective potential methods and it has been ignored by everyone since then. Coleman & Weinberg

(1973) observed that when ϕ developed a non-trivial minimum at $\phi = \Lambda$, their loop expansion failed for $\phi \leq \Lambda$. Therefore, the position, depth and even the existence of the minimum could not be trusted. In fact, some years earlier, Symanzik (1970) constructed a proof that $V(\phi)$ is always concave upwards. So, these new results (Callaway & Maloof 1983; Haymaker & Perez-Mercader 1983) show that the difficulty with $V(\phi)$ is not an artefact of Coleman & Weinberg's methods and they agree with the general result of Symanzik.

There may be some other way to do what Professor Kibble wants to do, but calculating $V(\phi)$ does not seem to be the way. Questions as to true and false vacua possibly cannot be decided purely within the context of quantum field theory and some additional physics might be needed.

References

Callaway, D. J. E. & Maloof, D. J. 1983 *Phys. Rev.* D **25**, 406.
Coleman, S. & Weinberg, E. 1973 *Phys. Rev.* D **7**, 1888.
Haymaker, R. W. & Perez-Mercader, J. 1983 *Phys. Rev.* D **27**, 1948.
Symanzik, K. 1970 *Communs math. Phys.* **16**, 48.

The cosmological constant

By S. W. Hawking, F.R.S.

University of Cambridge, Department of Applied Mathematics and Theoretical Physics, Silver Street, Cambridge CB3 9EW, U.K.

The cosmological constant is the quantity in physics that is most accurately measured to be zero: observations of departures from the Hubble law by distant galaxies place an upper limit of the order of 10^{-120} in dimensionless units. On the other hand, the various symmetry breaking mechanisms that we believe are operating in the Universe would give an effective cosmological constant many orders of magnitude larger, unless they are incredibly finely balanced. One answer would be to appeal to the anthropic principle, but a more attractive possibility is that there is a phase transition in $N = 8$ supergravity to a foam-like state which breaks supersymmetry and which appears flat on scales larger than the Planck length.

The history of the cosmological constant goes back to the seventeenth century. Newton realized that the so-called 'fixed stars' were like the Sun and he postulated that they were distributed approximately uniformly throughout space. This raised a problem: how could they remain at roughly constant distances from each other if they were attracting each other according to his law of gravity? Why did they not all fall together? One possible solution was to add a repulsive (in both senses of the word) 'cosmological' term to the Newtonian equation:

$$\nabla^2 \phi + \Lambda = 4\pi G \rho. \tag{1}$$

This would not make a significant difference to orbits in the solar system but it would allow static cosmological solutions with a uniform, non-zero average density. These solutions were unstable, however, because, if the density were slightly higher than average in some region, the gravitational attraction would dominate over the repulsion produced by the Λ term and the region would contract. Similarly, if the density were slightly less than average, the repulsion would win and would cause the region to expand indefinitely.

A static universe was still the accepted cosmological model when Einstein proposed the general theory of relativity in 1915. He therefore added a similar cosmological term to his field equation:

$$R_{ab} + \tfrac{1}{2} g_{ab} R + \Lambda g_{ab} = 8\pi G T_{ab}, \tag{2}$$

(I shall use units in which $c = 1$). This again allowed static cosmological solutions with a uniform non-zero density although, as before, they were unstable. However, a few years later, it was discovered that other galaxies were receding from our own with a velocity that increased with their distance from us. The static model of the Universe therefore had to be abandoned in favour of an expanding one. This removed the original reason for introducing the cosmological term. If one now regarded it as an adjustable parameter in the theory, one could attempt to measure it by observing departures from linearity in the magnitude redshift plot for distant galaxies (Sandage & Tammann 1982), figure 1.

If one assumes that the Universe is spatially homogeneous and isotropic (which seems to be a good approximation on large scales), then it is described by a Robertson–Walker metric with scale factor R. The Einstein equations give

$$3H^2 = 3\dot{R}^2/R^2 = 8\pi G\mu + \Lambda - k/R^2, \tag{3}$$

$$q_0 H^2 = \ddot{R}/R = -\tfrac{4}{3}\pi G(\mu + 3p) + \Lambda, \tag{4}$$

where μ is the average energy density and p is the average pressure. One can determine the Hubble constant H from the magnitude-redshift plot if one knows the intrinsic luminosity of a typical galaxy. One finds a value for H between 10^{-10}/year and 5×10^{-11}/year. If one assumed that all the galaxies had the same intrinsic luminosity, one could determine q_0 by measuring departures from linearity in the magnitude-redshift plot. One would expect however that the luminosities of the galaxies would have changed with time but one does not know how much. Nevertheless, one can be confident that $-5 < q_0 < 5$. One can also place a lower bound of 3×10^{27} cm on the scale factor R from the fact that one sees distant galaxies which are apparently randomly distributed. If one now uses the fact that the average pressure p in the universe is small, one can place an upper bound on $|\Lambda|$ of 10^{-54} cm^{-2}.

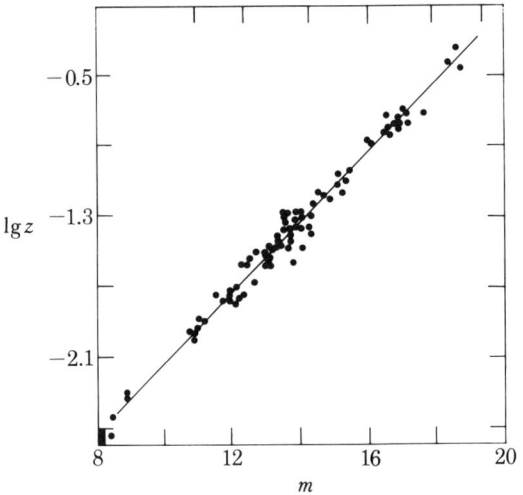

FIGURE 1. The magnitude redshift plot for the brighest galaxies in clusters of galaxies (see Sandage & Tamman 1982).

From the point of view of the classical general relativity, Λ is an *ad hoc* addition to the theory and these is no natural length scale with which to compare it. Thus, there is no reason to give it any particular value other than zero. However, when one takes quantum mechanics into account, the gravitational constant G has dimensions of mass^2 or length^{-2}. In other words, one can use units in which $\hbar = 1$ and one can write $G = M_P^{-2} = l_P^2$, where M_P is the Planck mass *ca.* 10^{19} GeV and l_P is the Planck length *ca.* 10^{-33} cm. One can then express the observational upper bound in the dimensionless form

$$\Lambda/M_P^2 < 10^{-120}. \tag{5}$$

This makes Λ the quantity in physics that is most accurately measured to be zero. By contrast, the observational upper limit on the mass of the photon obtained from spacecraft measurement of the Earth's magnetic field is only

$$m_\gamma^2/m_e^2 < 10^{-48}. \tag{6}$$

Yet we do not regard the mass of the photon as an adjustable parameter which just happens to have a very low value. Rather, we invoke a physical principle, gauge invariance, to make it exactly zero. The trouble is that there does not seem to be any similar principle or symmetry to which one can appeal to make Λ zero. In fact there seem to be several effects which would be expected to give rise to an effective value of Λ much larger than the observational upper limit (5) unless they cancel each other out to a very high degree of accuracy.

In a quantum state, such as the vacuum state, which is approximately invariant under local Lorentz transformations, the expectation value of the energy momentum tensor must be proportional to the metric. This gives rise to an effective cosmological constant Λ:

$$\Lambda = 8\pi M_P^{-2} \langle T_{00} \rangle. \tag{7}$$

In other words, the vacuum energy produces an effective cosmological constant. There are a number of effects that will contribute to the vacuum energy. To start with, the zero point fluctuations in each mode will contribute $\pm \frac{1}{2}\omega$ where the plus sign is for bosons and the minus sign is for fermions. Unless the number of boson and fermion degrees of freedom were equal (a possibility which I will come to), one would have to cut off the fluctuations at some frequency ω_0. One would then have an effective Λ of the order of $\omega_0^4 M_P^{-2}$. The most natural value of the cut off would be $\omega_0 = M_P$. This would violate the upper bound (5) by 120 orders of magnitude.

Even if the vacuum fluctuations were renormalized to zero, one would still get large changes in the vacuum energy when symmetries were broken. The contribution to Λ would be of the order of $\mu^4 M_P^{-2}$ where μ is the energy at which the symmetry is broken. We have good experimental evidence that chiral symmetry is broken around 200 MeV and that electro-weak symmetry is broken around 100 GeV. In addition there are indications that there may be a grand unified symmetry between strong and electroweak interactions which is broken somewhere between 10^{15} and 10^{19} GeV. It is very difficult to believe that the zero in the renormalization of the vacuum energy could have been chosen so exactly as to cancel out the contributions of all the symmetry breakings to better than one part in 10^{40}.

There is another possible symmetry apart from those mentioned above. It is supersymmetry, which relates fields of different spin. In a supersymmetric theory the number of boson and fermion degrees of freedom are equal so the infinite contributions to Λ cancel out. However, supersymmetry would also imply that all particles have the same mass, and clearly they don't. Thus, if the fundamental theory is supersymmetric, the supersymmetry must be broken at some energy greater than about 1000 GeV. This would give a positive contribution to Λ which would exceed the observational limit by 56 orders of magnitude. It might be possible to balance this with a negative contribution from some other form of symmetry breaking but this would require exceptionally fine tuning.

One possible explanation of the smallness of the cosmological constant would be to invoke the anthropic principle. One could imagine that there were very many different universes with different values of the cosmological constant. Only in those in which the cosmological constant was very small would it be possible for intelligent life to develop and ask the question: why is the cosmological constant so small? This idea will be discussed further in the lecture by Brandon Carter. Here, I will describe a different possibility, namely, that there is a phase transition to a state in which space–time is 'foam-like' on scales of the Planck length but appears smooth and nearly flat with zero cosmological constant on larger scales.

I shall assume that the fundamental theory of the Universe is $N = 8$ gauged extended supergravity (de Wit & Nicolai 1981). The reasons for this are that, first, supergravity theories seem to offer the only hope of quantizing gravity without non-renormalizable divergences. It is known that the $N = 8$ theory is finite at the one and two loop level (apart from a possible topological divergence that I shall describe below). There is a possible counter term with the right symmetry at the three loop level but it is not known whether it actually appears with a non-zero coefficient. If it does, it seems that one will have to go to a higher derivative theory with all the problems that has with ghost states: lack of unitarity and runaway solutions. The other attraction of supergravity theories is that they seem to provide the only way of unifying gravity with the other interactions of physics such as the strong and weak forces and electromagnetism. The $N = 8$ gauged theory is the largest of all such supergravity theories and the most likely to be finite at all numbers of loops. On the other hand, there seem to be several obstacles to accepting $N = 8$ gauged supergravity as the ultimate theory of the observed Universe.

(i) The ground state of the $N = 8$ theory is anti-de Sitter space which has a large negative cosmological constant Λ of order $-e^2 M_P^2$ where e is the coupling constant of the $SO(8)$ gauge fields.

(ii) The theory is supersymmetric whereas the observed Universe is not.

(iii) Although the $N = 8$ theory contains a large number of different particles, it does not seem to contain enough. In particular, the $SO(8)$ gauge group does not contain the $SU(3) \times SU(2) \times U(1)$ group which seems to be observed.

I shall show how a phase transition to a foam-like state could answer objections (i) and (ii). It might also answer (iii), though that remains to be determined.

Gauged extended supergravity theories for $N \geq 4$ contain scalar fields with a potential that has extremum points at negative values of the potential. One of these extrema gives rise to a supersymmetric solution which is anti-de Sitter space with constant values of the scalar fields. One might think that this solution would be unstable because the extremum point is a local maximum and the potential is unbounded below. Nevertheless, one can show that the solution is stable both to small fluctuations (Breitenlohner & Freedman 1982) and, indeed, to all fluctuations (Gibbons et al. 1982) provided that the fields obey certain boundary conditions (Hawking 1983). There are other extrema of the scalar potential (Warner 1983 a, b) that give rise to anti-de Sitter solutions with less supersymmetry or with no supersymmetry at all. However, the result of Gibbons et al. (1982) indicates that there cannot be any quantum tunnelling from the fully supersymmetric solution to these other solutions. One can understand this in the following way: the tunnelling would proceed by a 'bubble solution' which consisted of a region of the new solution surrounded by the fully supersymmetric solution. There would be an energy gain which would be proportional to the volume of the region and an energy loss which would be proportional to the area of the boundary of the region. In flat space one can make the ratio of volume to area indefinitely large by making the region large enough. Thus, it is always possible to tunnel. However, in anti-de Sitter space, the ratio of volume to area is bounded. Thus, it may not be possible to tunnel if certain inequalities are satisfied (Coleman & de Lucia 1980) and these inequalities hold in supersymmetric theories (Weinberg 1982).

The arguments given above indicate that it is not possible to break supersymmetry by tunnelling to one of the other extrema of the scalar potential. In $N \leq 4$ supergravity one can break supersymmetry by adding supermatter fields. One can even arrange it so that a positive contribution from the matter fields exactly cancels the negative cosmological constant of the gauged supergravity theory though this requires very fine tuning. However, there is no supermatter for

$N > 4$ so one cannot break $N = 8$ supersymmetry in this way. If one added $N = 4$ supermatter to $N = 8$ supergravity, one would obtain a theory that had at most four supersymmetries and that would be likely to have divergencies even at one loop. Thus the only possible way to break $N = 8$ supersymmetry without introducing unrenormalizable divergencies would seem to be to appeal to radiative corrections. However one would expect that there would be non-renormalization theorems which would prevent the breaking of supersymmetry by radiative corrections unless they were infinite. But, as was said before, the theory would be non-renormalizable if the radiative corrections were infinite. The one exception to this statement is a possible one-loop divergence that is proportional to a topological invariant.

The one-loop divergences of gauged supergravity theories have been analysed by Christensen et al. 1980 in the Euclidean régime, i.e. for positive definite metric. There are two possible terms:

$$A\chi + B\int \Lambda^2 \, dv, \tag{8}$$

where χ is the Euler number of the space. The coefficient B is zero for $N \geqslant 5$ (cf. Allen & Davis 1983). This is related to the nonexistence of supermatter for $N > 4$. The coefficient A is $(3-N)$ for $N \geqslant 3$. There is an alternative formulation of the ungauged $N = 8$ theory which is obtained by dimensional reduction from the $N = 1$ theory in eleven dimensions (Cremmer & Julia 1979). In this, 7 of the 70 scalar fields are replaced by antisymmetric tensor fields that have the same number of physical degrees of freedom. The coefficient A for this theory is zero (Siegel 1981; Duff 1981). However, this theory cannot be invariant under the group SO(8) because 7 is not a representation of SO(8). In fact it seems that it is a version of the ungauged $N = 7$ theory which has the same physical degrees of freedom as the $N = 8$ theory. It seems that in any version of the $N = 8$ theory which has SO(8) as a gauge group, the coefficient A will equal minus five, and certainly this is true of the only gauged $N = 8$ theory that we have, the one given by de Wit & Nicolai (1981).

The fact that A is non-zero means that one will have to add a term $-k\chi$ to the effective action where k is a scale dependent topological coupling constant. The fact that A is negative implies that topological fluctuations will be more important on small scales (the opposite of asymptotic freedom):

$$k = k_0 - 5\ln(l/l_P), \tag{9}$$

where l is a typical length scale of the topological fluctuation. At first sight, it would seem that the metric would have more and more topological fluctuations on smaller and smaller length scales and that space–time would not be a smooth manifold but would be something like a fractal. We do not know how to formulate such a theory and, even if we did, there would be no reason why space–time should have four dimensions rather than three and a half or π. Perhaps fortunately, it seems that there is a self-limiting effect which suppresses fluctuations on scales less than the Planck length. I shall give a brief outline of the idea here and more detail in another publication.

The Euclidean action for the gravitational field is not bounded below because the kinetic term for the conformal degree of freedom is negative definite (Gibbons et al. 1978). In order to deal with this, one has to divide the metrics in the path integral into conformal equivalence classes and perform the path integral over the conformal factor Ω along a contour in the complex Ω plane which crosses the real axis (cf. Hartle & Hawking 1983). One can deform the Ω contour so that it crosses the real axis at the lowest saddle point, i.e. the maximum value of the Euclidean action on the real Ω axis (there will always be a maximum if the cosmological constant is negative). One would expect that the dominant contribution to the path integral over the conformal factor would come from the saddle point.

Consider a metric which has a roughly uniform distribution of topological fluctuations. Under a constant conformal transformation Ω the action of the non-gravitational fields will be unchanged and the Euclidean action I of the gravitational field of a region of this metric will be:

$$I = C_1 \Omega^2 - C_2 \Omega^4 + C_3 \ln(\Omega/\Omega_0). \tag{10}$$

The first term is the gravitational action $-(16\pi)^{-1} M_P^2 \int R \, dv$. The constant C_1 will be roughly proportional to the square root of the Euler number χ of the region (Hawking 1978). The second term is the contribution of the cosmological constant. The constant C_2 will therefore be proportional to e^2. The third term is the contribution of the topological term in the action. The constant C_3 will be proportional to the Euler number χ and Ω_0 is related to the renormalization constant k_0.

If one neglects the topological term, the action will have a positive maximum value on the real Ω axis proportional to χe^{-2}. Thus, topological fluctuations will have higher actions and so will be suppressed. However, if one includes the topological term, the maximum value will occur for

$$\Omega^2 \approx C_4 e^{-2} \chi^{\frac{1}{2}} (1 + C_5 e), \tag{11}$$

where C_4 and C_5 are constant. If e is small, the maximum value of I will still be positive and proportional to χe^{-2}. Thus, topological fluctuations would still be suppressed. However, if e were greater than some critical value e_0, the maximum value of I would be negative and proportional to χ times some function of e. In this case, it would be more favourable to have topological fluctuations than not to have them and they would occur predominantly on the scale given by (11), i.e. about one topological fluctuation or 'bubble' per Planck volume. It therefore seems that the $N = 8$ theory may have two different régimes or phases: if $e < e_0$, the ground state is anti-de Sitter space and supersymmetry is unbroken; but if $e > e_0$, the ground state will contain a large number of topological fluctuations, rather like the confinement phase of Yang–Mills theory. There is, however, a difference from Yang–Mills theory: the coupling constant e is not renormalized and is scale independent. One would therefore expect the $N = 8$ theory to be either in the anti-de Sitter phase or in the space–time foam phase on all length scales according to whether e is less or greater than the critical value.

What would the space–time foam phase look like? To start with, it would almost certainly break supersymmetry because it is only a very few special spaces, like anti-de Sitter space, which possess Killing spinors, i.e. supercovariantly constant spinor fields. However, it would seem that it would break supersymmetry without giving a mass to the spin $\frac{3}{2}$ fields as would normally happen. Equation (11) indicates that the topological fluctuations or bubbles would occur predominantly on the Planck length scale. Thus, one might expect that on larger scales, space–time would appear smooth. However, the presence of the bubbles would mean that the apparent curvature on large scales could be very different from that on small scales. One can see this effect by considering solutions of the Einstein equations with a negative cosmological constant. A theorem of Yau (1977) implies that there exist large numbers of such solutions with arbitrarily high Euler number. In these, the volume integral of the square of the Weyl tensor will be of the order of χ. The Weyl tensor will be randomly orientated on the scale of the bubble length and will introduce shear, and hence convergence, on congruencies of geodesics. On a scale large compared with the bubbles it will therefore appear to have the same effect as a positive contricution to the Ricci tensor of order $\rho^{\frac{1}{2}}$ where ρ is the Euler number density per unit volume. Thus, the bubbles will tend to cancel out the negative Ricci tensor, which arises from the potential of the scalar fields. But why should they cancel it out to one part in 10^{120}? The answer is that, if it is favourable to have bubbles, then it is more favourable to have more bubbles. Thus, the density of bubbles tries to be as high as it can. However, if the apparent Ricci tensor on large scales becomes positive, then the whole space curls up on itself and has a small euclidean four-volume. In this case, the total number of bubbles would be limited. The most favoured case therefore is when the density of bubbles is such that the apparent large scale Ricci tensor is zero, i.e the apparent cosmological constant is zero.

The suggestion is therefore that the $N = 8$ gauged supergravity theory has a phase transition at a certain critical value of the gauged coupling constant e. Above that value, the ground state would be 'foam-life' with topological fluctuations on the Planck length scale but smooth and exactly flat on larger length scales. This bubble mechanism seems to be the only way of getting a zero effective cosmological constant without fine tuning. The foam would severely affect the propagation of elementary scalar particles but not that of low energy particles of spin half or higher (Hawking et al. 1980). The number of spin $\frac{1}{2}$ particles in the $N = 8$ theory is 56, which is large enough to accommodate all the observed particles. The observed spin 1 particles could not be the fundamental $N = 8$ particles because the group $SO(8)$ does not contain the observed $SU(3) \times SU(2) \times U(1)$. Presumably, they would have to be composites of the fundamental spin 0 or $\frac{1}{2}$ particles. Such composite particles would probably not be much affected by the space-time foam.

References

Allen, B. & Davis, S. 1983 *Physics Lett.* B **124** B, 353.
Breiltenlohner, P. & Freedman, D. Z. 1982 *Ann. Phys.* **144**, 249.
Christensen, S. M., Duff, M. J., Gibbons, G. W. & Rocek, M. 1980 *Phys. Rev. Lett.* **45**, 161.
Coleman, S. & de Lucia, F. 1980 *Phys. Rev.* D **21**, 3305.
Cremmer, E. & Julia, B. 1979 *Nucl. Phys.* B **159**, 141.
de Wit, B. & Nicolai, H. 1981 *Physics Lett.* B **108**, 285.
Duff, M. J. 1981 Antisymmetric tensors and supergravity. In *Superspace and supergravity* (ed. S. W. Hawking & M. Rocek), pp. 381–402. Cambridge: University Press.
Gibbons, G. W., Hawking, S. W. & Petty, M. J. 1978 *Nucl. Phys.* B **138**, 141.
Gibbons, G. W., Hull, C. M. & Warner, N. P. 1982 The stability of gauged supergravity. (DAMTP Preprint).
Hartle, J. B. & Hawking, S. W. 1983 Wave function of the Universe. *Phys. Rev.* D (Submitted.)
Hawking, S. W., Page, D. N. & Pope, C. N. 1980 *Nucl. Phys.* B **170**, 283.
Hawking, S. W. 1978 *Nucl. Phys.* B **144**, 349.
Hawking, S. W. 1983 The boundary conditions in gauged supergravity. *Physics Lett.* B (Submitted.)
Sandage, A. R. & Tammann, G. A. 1982 H_0, q_0 and the local velocity field. In *Astrophysical cosmology* (ed. H. A. Bruck, G. V. Coyne & M. S. Longair). Pontificae Academiae Scientiarum Scripta Varia. **48**.
Siegel, W. 1981 The wraiths of graphs. In *Superspace and supergravity* (ed. S. W. Hawking & M. Rocek), pp. 151–164. Cambridge: University Press.
Warner, N. P. 1983a Some properties of the scalar potential in gauged supergravity theories. California Institute of Technology preprint.
Warner, N. P. 1983b Some new extrema of the scalar potential of gauged N = 8 supergravity. California Institute of Technology preprint.
Weinberg, S. 1982 *Phys. Rev. Lett.* **48**, 1776.
Yau, S. T. 1977 *Proc. natn. Acad. Sci. U.S.A.* **74**, 1798.

Discussion

H. B. NIELSEN (*Bohr Institute, Copenhagen, Denmark*). Is it essential for one's understanding of the zeroness of the cosmological constant to have the $N = 8$ supergravity? The bubbles might occur in many gravity models. What are the essential assumptions?

S. W. HAWKING. $N = 8$ supergravity is not essential. But it is necessary to the situation that I have described, in that the small scale value of the cosmological constant is negative because bubbles always add a positive contribution to the apparent large scale cosmological constant. In fact one can obtain a zero effective cosmological constant without appealing to topological fluctuations at all: one can replace the density of bubbles by a 3-index anti-symmetric tensor field. This has no dynamics: the field equations are that the curl is constant. This provides a constant effective contribution to the cosmological constant and a similar argument to that for bubbles shows that the probability is highest when the total cosmological constant is zero: if it is negative, the action is positive and the probability is exponentially small, if it is positive, the Euclidean volume of the space is limited and the action is bounded below. The most favourable case is when the total cosmological constant is exactly zero.

Phil. Trans. R. Soc. Lond. A **310**, 311–322 (1983)
Printed in Great Britain

Large numbers and ratios in astrophysics and cosmology

By M. J. Rees, F.R.S.
Institute of Astronomy, Madingley Road, Cambridge CB3 0HA, U.K.

The masses and lifetimes of stars can be expressed in terms of fundamental constants. Such expressions always involve powers of the number $\hbar c/Gm_p^2$, whose huge magnitude stems from the weakness of gravity on microphysical scales. Our physical understanding of what determines *galactic* dimensions is not yet, however, on the same firm footing. Observational cosmology gives us three basic numbers that characterize our Universe; (i) the Robertson–Walker curvature radius (whose present-day value is $\gtrsim 10^{60}$ Planck lengths); (ii) the baryon-to-photon ratio (of order 10^{-9}); (iii) the amplitude of the initial metric fluctuations which triggered galaxy formation (of order 10^{-4}). We are unsure how (or, indeed, whether) these cosmological numbers can be derived from known physics.

1. Introduction: the significance of α_G

The feebleness of the gravitational force is expressed quantitatively by the value of the 'gravitational fine structure constant' $\alpha_G = Gm_p^2/\hbar c \approx 10^{-38}$ (or by other numbers related to α_G by factors like $\alpha_f = e^2/\hbar c$ or m_e/m_p, where m_p is the proton mass). Familiar arguments, summarized in the accompanying paper by Press & Lightman (1983), tell us that stars - whether they are main sequence stars (gravitationally bound fusion reactors) or white dwarfs - have masses of order the Chandrasekhar mass, $\alpha_G^{-3/2} m_p$ (or $\alpha_G^{-1} M_P$, where M_P is the Planck mass). Stars contain as many as 10^{57} baryons, because this is the number needed for gravitational binding energy to compete with thermal or degeneracy pressure.

Straightforward arguments show, furthermore, that stars are *long lived* (as well as very massive) because gravity is weak. An upper limit to stellar luminosities is the so-called 'Eddington luminosity' $L_E = 4\pi G m_p M/c\sigma_T$, where σ_T is the Thomson cross section. This is the luminosity for which radiation pressure on free electrons would balance gravity. The lifetime of a star can then be written as

$$t_* = Mc^2/L_E \times (\text{efficiency of rest-mass conversion}) \times (L_E/L_*), \qquad (1)$$

which can be expressed as

$$t_* = \tfrac{2}{3}(\alpha_f m_p/m_e)(e^2/m_e c^2)\alpha_G^{-1} \times \begin{pmatrix}\text{efficiency of rest}\\ \text{mass conversion}\end{pmatrix} \times L_E/L_*. \qquad (2)$$

The first term in brackets is of order 10; the efficiency of nuclear burning is $\lesssim 0.01$, and the stellar luminosity L_* is less than L_E because of other opacity additional to Thomson scattering, and because radiation provides only part of the total pressure. The important feature of (2) is that α_G^{-1} enters explicitly: the fiducial timescale determining the lifetime of stars – the time any object would take to lose its entire rest mass if it radiated with a luminosity L_E – is α_G^{-1} times the light travel time across a classical electron (or $\alpha_G^{-3/2} t_P$, where the Planck timescale, $t_P = (G\hbar/c^5)^{1/2} \approx 5 \times 10^{-44}$ s).

Figure 1 summarizes the physics of stars, planets, etc., in a mass–radius plot (see Press & Lightman 1983 for fuller discussion and references). The most striking feature is that significant phenomena occur for masses related to m_p by simple powers of α_G. The Planck mass, for which Compton and Schwarzschild radii are equal, is $\alpha_G^{-\frac{1}{2}} m_p$. A mass $\alpha_G^{-1} m_p$ corresponds to a black hole whose radius is the size of a proton: such a hole has Hawking (1975) temperature $kT \approx m_p c^2$, and radiates in a time of order $\alpha_G^{-\frac{3}{2}} t_P$ (i.e. a stellar lifetime, t_*). Stellar masses are of order $\alpha_G^{-\frac{3}{2}} m_p$. The mass scale $\alpha_G^{-2} m_p$ is also of significance as being the mass within a Hubble volume for a flat

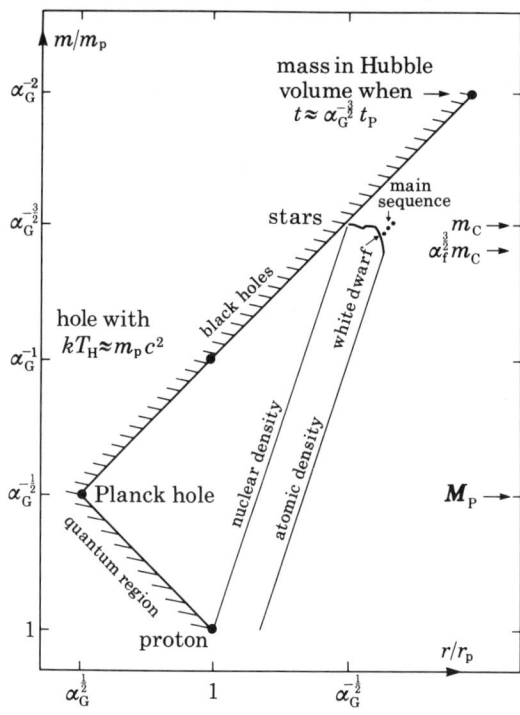

FIGURE 1. This diagram shows schematically, on a (logarithmic) mass–radius plot, how various characteristic scales of structure where gravitation is important involve simple powers of the 'gravitational fine structure constant' α_G. It is because α_G^{-1} is such a vast number, ca. 1.7×10^{38}, that so many powers of ten separate the astronomical and cosmological scales from the microphysical scale. Note, however, that the general shape of this diagram is insensitive to the value of α_G: if α_G^{-1} were somewhat smaller, the diagram would be altered only insofar as the separation between the atomic density and nuclear density lines (($\alpha_f^{-1} m_p/m_e$) horizontally) would look relatively larger.

Friedmann universe whose age is of order t_*. (Galactic masses can perhaps be included on this type of diagram, but as discussed in §4.1 the physics is less clearcut.) This diagram emphasizes that it is only because α_G^{-1} is so huge that so many powers of ten separate the macrophysical from the astrophysical scales. Nothing in the diagram, however, depends sensitively on the actual value of α_G^{-1} (which is 1.7×10^{38}). If it were (say) of order 10^{32} rather than of order 10^{38}, one could envisage a small-scale speeded up universe where H-burning stars still existed, but were less massive by ca. 10^9 and had lifetimes shorter by ca. 10^6.

A prime aim of cosmologists is to understand the scale and homogeneity of our Universe. There are around 10^{80} baryons and around 10^{89} thermal photons within the 'Hubble volume'. A universe must plainly have a large space–time volume for stars (and other entities of astrophysical interest) to form and evolve within it, but one would hope to account for this as an outcome of natural initial conditions.

A universe in which stars can form and evolve must persist for a time $t \gtrsim \alpha_G^{-\frac{3}{2}} t_P$. Unless the density is orders of magnitude below 'critical', or unless the mass–energy is overwhelmingly in some non-baryonic form (and it would be hard to form stars at all if either of these were the case) then the total number of baryons must be greater than about α_G^{-2}.

Independently of the properties of stars, we can note that a hot Friedmann cosmological model will not permit small-scale departures from thermodynamic equilibrium until it cools to the stage when the free electrons combine to form atomic hydrogen. This happens at a decoupling temperature T such that $kT \approx (0.02)\alpha_f^2 m_e c^2$, where the numerical factor 0.02 arises because the much smaller phase-space volume for bound electrons delays the recombination until well below the temperature of 1 rydberg (equivalent to $0.5\alpha_f^2 m_e c^2$). The time t_{rec} taken to cool to this temperature in a radiation-dominated universe exceeds t_P by a factor of around $10^3 \alpha_G^{-1}(m_p/m_e)^2 \alpha_f^{-4}$: another very large number that explicitly involves α_G^{-1}.

The above considerations show – for what it is worth – that it is because α_G^{-1} is so large that a vast universe is a prerequisite for the existence of stars, cosmologists, and other manifestation of thermodynamic disequilibrium (see Carr & Rees 1979 for further discussion of these points).

2. ROBERTSON–WALKER CURVATURE AND THE DENSITY PARAMETER Ω

2.1. *Quantifying the curvature scale*

If the Universe had resembled a Friedmann model since t_P, a comoving scale initially equal to the Planck length would by now have grown by *ca.* 10^{30}, because the scale factor varies as $t^{\frac{1}{2}}$ for a radiation-dominated expression. But the present Hubble radius is *ca.* 10^{60} Planck lengths. The Robertson–Walker curvature radius, which is certainly not much smaller than the present Hubble radius, is thus $\gtrsim 10^{30}$ larger than the 'natural' scale. This problem – the so called 'flatness problem' – introduces another large number, and is perhaps solved by the concept of an 'inflationary' universe, as discussed in this volume by Kibble (1983).

We can write

$$\left\{\frac{\text{Robertson-Walker curvature radius}}{\text{Planck scale expanded to present epoch } t_0}\right\} = \left(\frac{t_0}{t_P}\right)^{\frac{1}{2}} \left(\frac{t_{\text{eq}}}{t_0}\right)^{\frac{1}{6}} \frac{1}{|\Omega - 1|^{\frac{1}{2}}}, \qquad (3)$$

where Ω is the density parameter, defined as ρ/ρ_c, the critical density being $\rho_c = (\frac{8}{3}\pi G t^2)^{-1}$. The extra factor $(t_{\text{eq}}/t_0)^{\frac{1}{6}}$, where t_{eq} is the time at which the expansion switches from being radiation dominated to being matter-dominated, arises because for $t > t_{\text{eq}}$ the scale factor grows as $t^{\frac{2}{3}}$ rather than as $t^{\frac{1}{2}}$.

The actual density of the Universe – and, in particular, the question of whether $\Omega = 1$ – is of crucial importance. We must consider all forms of matter: dark as well as luminous; non-baryonic as well as baryonic.

2.2. *Contributions to Ω*

2.2.1. *Luminous matter: galaxies and gas*

The present Hubble constant (not in any sense, of course, a 'fundamental' constant) is still uncertain, so in quoting numerical values I will introduce a quantity $h = (3 \times 10^{17} s/t_H)$; the experts advocate values of h in the range 0.5–1. See Sandage & Tammann (1982) and Hanes (1982) for recent assessments from differing viewpoints. The baryon density is then $n_b = 3 \times 10^{-6}(\Omega_b h^2)$ cm^{-3}, where Ω_b is defined as the fraction of the critical density in baryonic form. An important number which *is* perhaps of fundamental significance is the ratio of the baryon density to the density of photons in the microwave background, apparently a black body with $T \approx 2.7$ K. This is then

$$\mathscr{S}^{-1} = n_b/n_\gamma \approx 3 \times 10^{-8} (T/2.7\,\text{K})^{-3} (\Omega_b h^2), \tag{4}$$

$\mathscr{S}$ being a measure of the entropy per baryon.

A direct lower limit on Ω_b of order 10^{-2} can be set from the observed mean density of baryons in conspicuously 'luminous' form (visible galaxies, and intergalactic gas revealed by its X-ray emission), implying a baryon-to-photon ratio of not less than about $3 \times 10^{-10} h^2$. This is the number which GUT models must attempt to explain (see §3).

There is no firm evidence for any anti-matter in the Universe (apart from a small fraction of anti-particles in cosmic rays, which could have been produced in high-energy collisions). Strong constraints on the presence of anti-matter in and around our Galaxy are set by the measured limits to the γ-ray background. Nevertheless, if one is strictly agnostic and free from theoretical preconceptions, one can certainly envisage that the Universe might possess matter–anti-matter symmetry (i.e. that the overall net baryon number, and $\langle n_b/n_\gamma \rangle$, are zero) provided that the scale of the regions of each 'sign' is at least as large as a cluster of galaxies.

2.2.2. *Evidence for unseen mass*

There are dynamical indications of some unseen mass that may or may not be baryonic. The most convincing evidence comes from applications of the virial theorem to rich clusters of galaxies (e.g. the Coma cluster), from the dynamics of our local group of galaxies, and from the statistical technique known as the 'cosmic virial theorem', whereby one analyses the deviations from Hubble-law motions induced in galaxies by their neighbours. These studies are still bedevilled by observational problems, but they broadly suggest that Ω is in the range 0.1–0.2: in other words, there is perhaps ten times as much non-luminous matter as there is in stars and detectable gas. Ten times as much gravitating stuff is implicated in the relative motions of galaxies as in the internal dynamics of individual galaxies. The 'unseen' mass must be in diffuse halos around galaxies, or must pervade clusters or groups of galaxies. The evidence suggesting $\Omega = (0.1 - 0.2)$ comes from studying the dynamics of systems on scales $(1-2)\, h^{-1}$ Mpc.†

It is interesting to ask whether the data *permit* $\Omega = 1$, the value favoured by advocates of 'inflationary' cosmology. The short answer is that this is compatible with the data only if M/L continues to increase with length scale out to $\gtrsim 10 h^{-1}$ Mpc, so that on scales where the virial theorem can be reliably applied the unseen mass is less 'clumped' than the luminous mass. Much attention has been given to the velocity field in the local super-cluster, whose scale is 10–20 Mpc. Our infall velocity towards the Virgo cluster, relative to the mean Hubble flow, is apparently too small to permit $\Omega = 1$ *if* the total mass throughout the supercluster is distributed like the galaxies. However, if one drops this assumption, the results become quite inconclusive (Hoffmann & Salpeter 1982).

Only 10 % (and maybe as little as 1 %) of the mass–energy of the Universe is thus in 'known' form. All that can confidently be said about the unseen mass is that it is 'dark': it has a much higher mass-to-light ratio than the ordinary luminous content of galaxies. Whereas the inner parts of typical galaxies have M/L which is $\lesssim 10$ times the solar mass-to-light ratio, if the unseen mass contributes a density parameter Ω it must have M/L exceeding $2300\,\Omega h$ solar units. Possible

† 1 pc $\approx 3 \times 10^{16}$ m.

forms of unseen mass are discussed elsewhere (see, for example, Einasto & Rees 1983) so I will give just a brief summary here.

2.2.3. Baryonic forms for the unseen mass

If galaxies and clusters were 'assembled' from sub-units that condensed earlier, most of the initial baryons might have been incorporated in a pre-galactic population of stars: these stars, or their remnants, could perhaps now have a high M/L and contribute to the unseen mass. Ideally, one would like to be able to calculate what happens when a cloud of 10^6–$10^8 M_\odot$ condenses out soon after recombination: does it form one (or a few) supermassive objects, or does fragmentation proceed efficiently down to low-mass stars? Our poor understanding of what determines the masss spectrum of stars forming now (in, for instance, the Orion nebula), gives us little confidence that we can calculate the nature of pregalactic stars, born in an environment very different from our (present-day) Galaxy.

Although we cannot confidently predict what these pregalactic stars would be like (Kashlinsky & Rees 1983), there are several constraints which, in combination, imply that if there are enough of them to provide the unseen mass, the individual masses must either be less than $0.1 M_\odot$ or else in the range 10^3–$10^6 M_\odot$. Masses above $ca.$ $0.1 M_\odot$ would contribute too much background light unless they had all evolved and died, leaving dark remnants. But the remnants of ordinary massive stars of 10–$100 M_\odot$ would produce too much material in the form of heavy elements. Limits on the range 100–$1000 M_\odot$ are uncertain because only ^{4}He may be ejected, the 'heavies' in the core collapsing into a black hole remnant. An uncertainty in the evolution of massive or supermassive stars is the amount of loss during H-burning; however the hypothesis that most mass goes into very massive objects (v.m.os) of greater than about $10^3 M_\odot$ is compatible with the nucleosynthesis constraints. A further consideration favouring these high masses is that v.m.os are likely to terminate their evolution by a collapse which swallows most of the mass: if most of the material were ejected, 'recycling' through several generations would be necessary in order to end up with most of the material in black holes rather than gas. Detailed discussions of pregalactic stars are given by Carr et al. (1983) and by Tarbet & Rowan–Robinson (1982).

2.2.4. Non-baryonic unseen mass?

If neutrinos have negligible rest mass, the present density expected for relic neutrinos from the big bang is $n_\nu = 110\,(T_\gamma/2.7\,\text{K})^3\,\text{cm}^{-3}$ for each two-component species. This conclusion holds for non-zero masses, provided that $m_\nu c^2$ is far below the thermal energy ($ca.$ 5 MeV) at which neutrinos decoupled from other species and that the neutrinos are stable for the Hubble time. Comparison with the baryon density shows that neutrinos outnumber baryons by such a big factor ($ca.$ $\mathscr{S}$) that they can be dynamically dominant over baryons even if their masses are only a few electron volts. In fact, a single species of neutrino would yield a contribution to Ω of $\Omega_\nu = 0.01\,h^{-2}(m_\nu)_{\text{eV}}$, so if $h = 0.5$, only $25\,\text{eV}$ is sufficient to provide the critical density.

The entire range $100\,h^2\,\text{eV}$–$3\,\text{GeV}$ is incompatible with the hot big bang model (Gunn et al. 1978). (For $m_\nu > 3\,\text{GeV}$, the rest mass term in the Boltzmann factor would kill off most of the neutrinos before they decoupled; the number surviving would be less than $ca.$ n_b.) If any species of neutrino were discovered to have a mass in this excluded range, it would show that one cannot extrapolate the hot big bang back to $kT \gtrsim 5\,\text{MeV}$, and that most of the photons must have been generated at later times.

(Such arguments are familiarly expressed by saying that the hypothetical particles of non-zero rest mass would 'close the Universe by a large factor'. This loose phrase is in fact rather misleading. The geometry of the Universe is likely to have been laid down at very early stages by mechanisms that do not 'know' what the dominant constituents will be after 10^{10} years. If the Universe were indeed 'flat' (e.g. for 'inflationary' reasons), it would expand with $\Omega = 1$ (i.e. with a density of $(\frac{8}{3}\pi G t^2)^{-1}$) for all t. If the neutrinos were in the excluded mass range, or if there were, for instance, too many primordial monopoles, the observational incompatibility would be that the baryonic fraction of the total density would, for $t \approx 10^{10}$ years, be much less than 10^{-2}.)

Physicists have other particles 'in reserve' – right handed neutrinos, photinos or gravitinos, for instance – which (if they existed) could have been in thermal equilibrium with other species at very early times, and therefore contribute to Ω in an analogous way. The only difference would be that (n/n_γ) could be less than for neutrinos because the other '...inos' may have decoupled before muons (or even hadron pairs) annihilated: the later annihilations would then boost the neutrinos but not the still more weakly coupled '...inos'.

Any such particles would be dynamically important not only for the expanding Universe as a whole but also for large bound systems such as clusters of galaxies. This is because they would now be moving slowly: if the Universe had cooled homogeneously, primordial neutrinos would now be moving at around 200 $(m_\nu)_{eV}^{-1}$ km s^{-1}. They would be influenced even by the weak ($ca. 10^{-5} c^2$) gravitational potential fluctuations of galaxies and clusters. If the three (or more) types of neutrinos have different masses, then the heaviest will obviously be gravitationally dominant, since the numbers of each species should be the same.

It was conjectured more than a decade ago (Cowsik & McLelland 1972; Marx & Szalay 1972) that neutrinos could provide the 'unseen' mass in galactic halos and clusters. In recent years, astrophysicists have explored this possibility in some detail, and considered scenarios for galaxy formation in which neutrino clustering and diffusion play a key role. These scenarios have several appealing features, even though they lead to some new problems.

We may inhabit a universe where, on the largest scales, the baryons are merely a tracer for the distribution of a gravitationally-dominant neutrino sea. Neutrinos of mass $ca. 10$ eV, gravitinos, monopoles or axions are just some of many candidates for the unseen mass in the Universe: as far as the astronomical evidence goes, the unseen mass could equally well be low mass stars, or black holes of up to at least 10^6 solar masses (which could be either primordial, or the remnants of a generation of very massive pregalactic stars). There is no lack of candidates for unseen mass, either baryonic or non-baryonic.

2.3. *Primordial nucleosynthesis: need for non-baryonic matter if $\Omega = 1$?*

Some considerations based on primordial nucleosynthesis seem to favour the non-baryonic option, especially if the total density corresponds to $\Omega = 1$. Primordial nucleosynthesis depends on two things: the expansion timescale at 0.1–1 MeV and the baryon density (which is proportional to $\mathscr{S}^{-1} \propto \Omega_b h^2$). The predicted ^{4}He abundance is rather insensitive to the matter density: for $\Omega_b h^2 \gtrsim 10^{-2}$ (corresponding to $\mathscr{S} \lesssim 3 \times 10^9$) the density of baryons is high enough to ensure that most of the neutrons that survive when the neutron–proton ratio 'freezes out' at $kT \approx 1$ MeV get incorporated in ^{4}He.

The cosmic helium abundance can however be measured with sufficient precision to suggest that the primordial ^{4}He is less than 26 % at the 3σ level (Pagel 1982). This is compatible with $\Omega_b h^2 \lesssim 0.1$ but probably not with $\Omega_b h^2 = 1$ (for ≥ 3 species of neutrinos). The strongest constraint on Ω_b from primordial nucleosynthesis comes, however, not from ^{4}He but from deuterium. This is an intermediate product in helium formation, the amount emerging from the big bang being a steeply decreasing function of Ω_b. Only if $\Omega_b h^2 < 0.025$ can the observed deuterium abundance be produced in a standard hot big bang. The strength of this constraint stems from the failure of astrophysicists in the last decade to suggest any other plausible way of making deuterium.

The combined arguments from primordial nucleosynthesis suggest that $\Omega_b h^2$ is in the range 0.01–0.025 (permitting a value of Ω_b no higher than 0.1, even for $h = 0.5$). See Schramm (1983) for a recent summary. If the lepton number for ν_e and $\bar{\nu}_e$ were non-zero, then the neutron–proton equilibrium ratio would be shifted, affecting ^{4}He production. It is thereby possible in principle to accommodate a higher Ω_b (David & Reeves 1980). However, in order to make much difference, the neutrino lepton number must be of order the photon number; that is, $\mathscr{S}$ times larger than the baryon number. There are other possible complications and 'escape clauses' (involving large-amplitude inhomogeneities in the baryon distribution, etc.). But these considerations suggest that, unless one is to abandon the standard hot big bang model completely, the idea of *non*-baryonic unseen mass is very appealing, especially if (for other reasons) one favours an overall cosmological density as high as the critical value ($\Omega = 1$). Because the relevant parameter in primordial nucleosynthesis is $n_b/n_\gamma \propto \Omega_b h^2$, more precise comparison of models with observation must await a firmer value of the Hubble constant. If $h = 1$ (corresponding to Hubble time of 10^{10} years) then the simplest inference would be that most of the unseen mass – both in the halos of individual galaxies and in clusters and groups – was non-baryonic; but if $h = \frac{1}{2}$ (corresponding to a Hubble time of 2×10^{10} years) the *lower* limit to $\Omega_b h^2$ set by the requirement not to overproduce D + ^{3}He implies that some unseen matter – maybe that in halos, if not in intergalactic space – is baryonic, though only enough to contribute $\Omega_b \approx 0.1$. It is remarkable that the simplest and least arbitrary from of hot big bang model can, for a suitable choice of Ω_b, account for D, ^{4}He and ^{7}Li (Yang et al. 1983).

The production of primordial helium is among the few cosmic phenomena sensitive to the actual value of the weak interaction coupling constant. The resultant ^{4}He abundance is $ca.$ 25 %, rather than $ca.$ 0 % or $ca.$ 100 %, because the reactions controlling the neutron–proton ratio ($p + e^- \rightarrow n + \nu$, $p + \bar{\nu} \rightarrow n + e^+$) freeze out when kT has dropped to a value of order $(\Delta m) c^2$, where Δm is the proton–neutron mass difference, so the Boltzmann factor is neither close to unity nor ultra-small. (Another process sensitive to neutrino cross sections is the explosion of a supernova; were these cross sections smaller, neutrinos would not be sufficiently well trapped in the stellar core, bouncing at neutron densities, to drive the shock wave which blows off the stellar envelope. If the weak interactions were a factor of 10 weaker, we would have a universe composed primarily of ^{4}He, where supernovae could not explode.)

3. THE BARYON–PHOTON RATIO

Recent ideas on baryon synthesis – if the grand unified theories (GUTs) on which they are based are borne out by future developments – might allow us to test whether the Friedmann models apply back at temperatures of around 10^{15} GeV, corresponding to times $ca.$ 10^{-36} s. The observed baryon-to-photon ratio (equation (4)), a measure of the fractional excess of baryons over their antiparticles at early times when $kT > m_p c^2$, is $ca.$ 10^{-9}. If this number were much smaller than $\alpha_G^{\frac{1}{2}}$, the Universe would not be baryon-dominated when its age was of order a characteristic stellar lifetime. The value of the net baryon excess arising from out-of-equilibrium decay of X and Y particles can be computed, given a specific GUT (Kolb & Wolfram 1980); it involves a small parameter related to the CP-violation parameter in weak interactions. This work is not yet on the same footing as the calculations of primordial helium and deuterium: it is perhaps at the same level as nucleosynthesis was in the pioneering days of Gamow and Lemaitre. But if it could be firmed up it would represent an extraordinary triumph. The mixture of radiation and matter characterizing our Universe would not be *ad hoc* but would be a consequence of the simplest initial conditions. Also, as well as vindicating a GUT, it would reassure us about extrapolating in one bound, based on a Friedmann model, right back to the threshold of classic cosmology, almost back to the Planck time. On a logarithmic scale, this is a bigger extrapolation from the nucleosynthesis era than is involved in going to that era from the present time. It would also place constraints on dissipative processes arising from viscosity, phase transitions, black hole evaporation, etc., which might occur as the Universe cooled through the 'desert' between 10^{15} and 100 GeV. Although these ideas are still speculative, the 'prediction' of the photon–baryon ratio may turn out to offer one of the few empirical tests of GUTs.

If the baryon–photon ratio could be calculated, this would determine Ω_b. If $\Omega_b < 1$, then a strictly flat universe would require some non-baryonic contribution. Of course, one may eventually have theoretical knowledge of the rest masses of all other relevant particles; such information, in conjunction with knowledge of n_b/n_γ, would determine their contribution to Ω also. Looked at from this point of view, it perhaps seems coincidental that non-baryonic matter should dominate, but only by an order of magnitude rather than a vastly larger factor.

4. The origin of galaxies: primordial fluctuations

4.1. *Galaxies: the basic units*

The basic units delineating the Universe's large scale structure are, of course the *galaxies*. Much is now understood about their morphology and internal dynamics. But we still do not know *why* galaxies exist, why the most conspicuous large-scale features of the cosmic scene should be these gravitating aggregates of 10^{10}–10^{12} stars, with dimensions 10^4–10^5 light years. Galaxies have a broad luminosity function, but are no less standardized than stars. Whereas we understand stellar masses, we do not, however, have a convincing commensurate expression for galactic masses. Even worse, we do not know whether the explanation we seek lies within the province of the astrophysicist or the cosmologist. Conceivably the right characteristic mass is somehow 'imprinted' in the early Universe; alternatively, the galactic mass may be singled out by physical processes, just as stars in the stable mass range $\alpha_G^{-1} M_P$ are the end-product of condensation from a broad mass-spectrum of inhomogeneities (in, for example, the Orion nebula) without having to be favoured in the interstellar medium.

An indication that galactic dimensions may be determined by astrophysical processes comes from the following considerations (cf. Rees & Ostriker, 1977; Silk 1977; Binney 1977). A protogalaxy may have started its life as a massive gas cloud, not yet fragmented into stars. Two timescales are important in determining how a self-gravitating gas cloud evolves. The first of these is the dynamical, or free-fall, time t_{dyn}: this is of order $(G\rho)^{-\frac{1}{2}}$, its precise value depending on the geometry of the collapse. The second is the radiative cooling timescale: this depends on the gas temperature T_g, and can be written $t_{\text{cool}} = T_g/\rho f(T_g)$, where $f(T_g)$ depends on the composition and ionization of the gas and can be calculated from atomic physics (the cooling rate per unit volume is proportional to $\rho^2 f(T)$, which is why t_{cool} has the quoted dependence on T_g and ρ).

If $t_{\text{cool}} > t_{\text{dyn}}$, a cloud of mass M and radius R can be in quasi-static equilibrium, with T_g equalling the virial temperature $kT_g = GMm_p/R$. But if $t_{\text{cool}} < t_{\text{dyn}}$ such equilibrium is impossible:

the cloud cools below the virial temperature, and undergoes free-fall collapse or fragmentation. The criterion $t_{\text{cool}} \lesssim t_{\text{dyn}}$ also determines whether a shock developing during infall is isothermal (permitting a density enhancement of order the Mach number squared) or adiabatic (with a density enhancement not exceeding 4). We would expect clouds to fragment only if they enter the part of the M–R plane where $t_{\text{cool}} < t_{\text{dyn}}$. A simple calculation shows that this happens below a mass-independent radius

$$R_c = (e^2/m_e c^2)\,[\alpha_G^{-1} \alpha_f^2 (m_p/m_e)^{\frac{1}{2}}] \approx 75\,\text{kpc}, \tag{5}$$

for all masses such that the virial temperature at this radius lies in the range $\alpha_f^2 m_e c^2 \lesssim kT_g \lesssim m_e c^2$ (i.e. for values of kT between a rydberg and the electron rest mass, when non-relativistic bremsstrahlung is the main radiative process). The mass for which the virial temperature at R_c is $kT_g = \alpha_f^2 m_e c^2$ is

$$M_c \approx [\alpha_G^{-2} \alpha_f^5 (m_p/m_e)^{\frac{1}{2}}]\,m_p \approx 10^{12} M_\odot. \tag{6}$$

Masses below M_c cool efficiently by H and He recombination and line emission, even at larger radii than R_c. Clouds with mass below M_c will readily fragment; but above M_c fragmentation is impossible unless the cloud contracts to $R < R_c$.

These masses and radii are of the general order relevant to large galaxies, and play a role in many schemes for galaxy formation. However some cosmological processes in the early Universe must have given rise to gas clouds spanning the range around M_c: only then can mass-dependent cooling processes single out a preferred scale for galaxies. Two points are nevertheless uncontroversial:

(1) the early Universe must have contained *some* inhomogeneities (despite its overall Friedmannian character); otherwise its baryon content would now, after *ca.* 10^{10} years, still just comprise uniform neutral H and He.

(2) Whereas gas dynamical and dissipative effects must have been important in the formation of individual galaxies (so that the characteristics of galaxies are related only indirectly to the form of the primordial fluctuation spectrum), any inhomogeneities on much larger scales must be induced primarily by gravitation. The clustering properties of galaxies are the best evidence we have on this. (Note, however, that we can use these to infer the total density fluctuations only insofar as galaxies are a good tracer for the overall mass distribution.)

4.2. *Initial metric fluctuations*

If metric fluctuations exist in the early Universe, then we can constrain their amplitude ϵ on various length scales l as indicated in figure 2. The formation of galaxies by the present epoch from initial curvature fluctuations requires a certain minimum amplitude, which depends on the type of fluctuations (adiabatic or isothermal?) and on the nature of the hidden mass. These *lower* limits representative of classes of models are shown on the diagram. When these models are analysed in detail, other characteristic masses emerge which may be relevant to galaxies, so that the considerations leading to (5) and (6) may not be the complete story. In particular, a significant role is played by the largest mass scale of which free-streaming motions can homogenise the non-baryonic particles (neutrinos?). This is essentially the mass within the particle horizon when kT drops to $m_\nu c^2$: its value is $M_P(M_P/m_\nu)^2$, the analogue of the Chandrasekhar mass for a gravitationally-bound neutrino cloud (with m_ν replacing m_p).

The upper limits in figure 2 are not stringent except for the scales probed by anisotropy measurements of the microwave background, but the following conclusions can be drawn:

(i) if ϵ follows a power law $\epsilon(l) \propto l^x$, then $x > -0.15$;

(ii) if ϵ is *independent of scale* (the Harrison (1970)–Zel'dovich (1972) spectrum) as expected in 'inflationary' models, then its value is pinned down in the range 10^{-4}–10^{-5};

but,

(iii) if ϵ does not follow a smooth power law, we have significant constraints only for 4 (out of around 30) powers of ten in l.

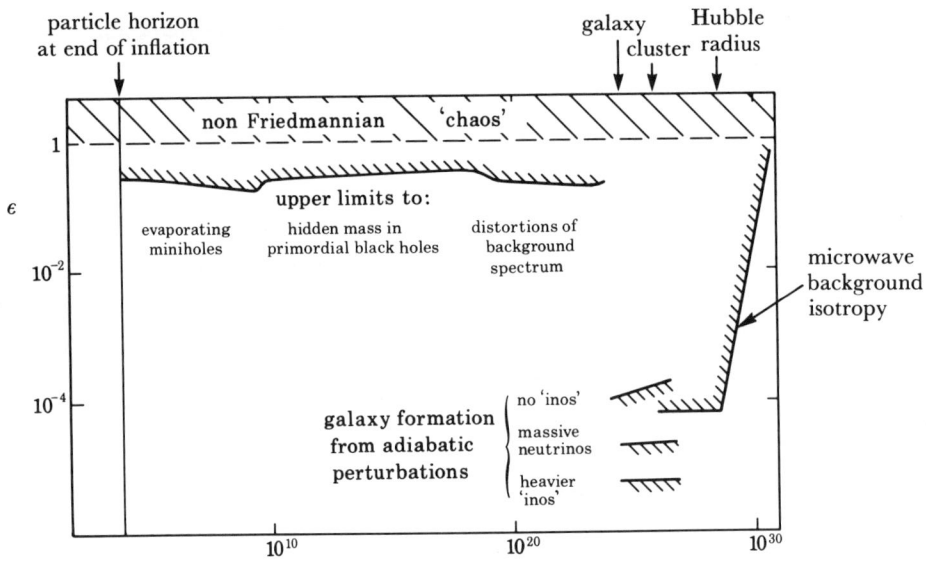

$l =$ (length scale/Planck length)$_\text{comoving coordinates}$

FIGURE 2. This diagram depicts the empirical limits on the amplitude ϵ of adiabatic metric perturbations on various scales. On large scales, the microwave background isotropy offers stringent upper limits. On smaller scales the limits are much less good. If bound systems (galaxies and clusters) in the present Universe evolved from adiabatic initial fluctuations, then *lower* limits are implied to the amplitude on the relevant scales. These limits depend somewhat on the detailed model, and on the nature of the unseen mass: three options are plotted. Because the various limits span a factor *ca.* 10^{30} in comoving length scale, these limits (rough though they are) constrain the slope of any power law $\epsilon \propto l^x$. If the Universe contains scale-independent metric fluctuations ($x = 0$), then the amplitude is pinned down to lie in the range 10^{-4}–10^{-5}. The diagram is drawn assuming a 'flat' background Universe with $\Omega = 1$, but only the large-scale limits are sensitive to this assumption.

5. Conclusions

The basic properties of stars can all be straightforwardly calculated: their large masses and timescales are a consequence of the vastness of the number $\alpha_G^{-1} \approx 1.7 \times 10^{38}$.

On the still larger scale of galaxies and the Universe, theories are much more provisional. Observational cosmology reveals three important constants.

(i) *The Robertson–Walker curvature radius.* The fact that the Universe is still expanding, after *ca.* $10^{60} t_P$, with a density within an order of magnitude of the 'critical' density, implies that the initial curvature radius at t_P (or at the end of an 'inflationary' phase) is more than *ca.* 10^{30} times larger than the horizon scale at that epoch. The precise value of this curvature radius depends on the density parameter Ω.

(ii) *The baryon-to-photon ratio.* This ratio $\mathscr{S}^{-1}$, given by (3), is a measure of the entropy per baryon. Grand unified theories suggest that it can be explained in terms of baryon non-conservation processes at $t \approx 10^{-36}$ s ($kT \approx 10^{15}$ GeV).

(iii) *The fluctuation amplitude.* The prime mystery is perhaps why the large scale Universe is so homogenous. However *some* fluctuations are essential in order to trigger galaxy formation. If these metric fluctuations have a scale-independent random-phase character, then the amplitude is pinned down to be $\epsilon \approx 10^{-4}\text{--}10^{-5}$, a number which theorists may hope to calculate; however if the fluctuations have a more general character, few constraints can be set.

The status of the galactic mass is still uncertain. It may be explicable in terms of physical processes at recent epochs; on the other hand, the scale of galaxies we see may be a consequence of fluctuations imprinted at early times.

Insofar as the aim of physics is to erode the number of independent underivable constants, it is gratifying that there is a serious chance of calculating the quantities listed above in terms of microphysical parameters.

References

Binney, J. J. 1977 *Astrophys. J.* **215**, 483.
Carr, B. J., Bond, J. R. & Arnett, W. D. 1983 *Astrophys. J.* (In the press.)
Carr, B. J. & Rees, M. J. 1979 *Nature, Lond.* **278**, 605.
Cowsik, R. & McLelland, J. 1972 *Phys. Rev. Lett.* **29**, 669.
David, Y. & Reeves, H. 1980 In *Physical cosmology* (ed. R. Balian, J. Audouze & D. N. Schramm), p. 443. Amsterdam: North-Holland.
Einasto, J. & Rees, M. J. 1983 *Nature, Lond.* (Submitted.)
Gunn, J. E., Lee, B. W., Lerche, I., Schramm, D. N. & Steigman, G. 1978 *Astrophys. J.* **223**, 1015.
Hanes, D. A. 1982 *Mon. Not. R. astr. Soc.* **201**, 145.
Harrison, E. R. 1970 *Phys. Rev.* D **1**, 2726.
Hawking, S. W. 1975 *Communs math. Phys.* **43**, 189.
Hoffmann, G. L. & Salpeter, E. E. 1982 *Astrophys. J.* **263**, 485.
Kashlinsky, A. & Rees, M. J. 1983 *Mon. Not. R. astr. Soc.* (In the press.)
Kibble, T. W. B. 1983 *Phil. Trans. R. Soc. Lond.* A **310**, 293.
Kolb, E. & Wolfram, S. 1980 *Nucl. Phys.* B **172**, 224.
Marx, G. & Szalay, A. S. 1972 *Proceedings of the Neutrino 72 Conference*, vol. I, p. 191. Budapest: Technoinform.
Pagel, B. E. J. 1982 *Phil. Trans. R. Soc. Lond.* A **307**, 19.
Press, W. H. & Lightman, A. P. 1983 *Phil. Trans. R. Soc. Lond.* A **310**, 323.
Rees, M. J. & Ostriker, J. P. 1977 *Mon. Not. R. astr. Soc.* **179**, 541.
Sandage, A. R. & Tammann, G. A. 1982 In *Astrophysical cosmology* (ed. H. A. Brück, G. V. Coyne & M. S. Longair), p. 23. Vatican Publications.
Schramm, D. N. 1983 *Phil. Trans. R. Soc. Lond.* A **307**, 43.
Silk, J. I. 1977 *Astrophys. J.* **211**, 638.
Tarbet, P. W. & Rowan-Robinson, M. 1982 *Nature, Lond.* **298**, 7.
Yang, J., Turner, M. S., Steigman, G., Schramm, D. N. & Olive, K. A. 1983 *Astrophys. J.* (Submitted.)
Zel'dovich, Y. B. 1972 *Mon. Not. R. astr. Soc.* **160**, 1P.

Discussion

W. H. McCrea, F.R.S. (*University of Sussex, Brighton BN1 9QH, U.K.*). If the amount of dark matter in the Universe greatly exceeds the amount of luminous (baryonic) matter, then, on the hypothesis that the dark matter is baryonic, the luminous matter would be only a small fraction of all the baryonic matter. The abundances of helium and deuterium quoted by Professor Rees are necessarily derived from observation of only that small fraction. Does Professor Rees consider that these abundances would have to be regarded as significant for all the rest? Otherwise, they would not be a compelling reason for rejecting the hypothesis that most of the mass of the Universe is in the form of baryonic matter.

M. J. Rees. The observed baryon content of the Universe could be atypical of all baryonic matter in its He and D abundance only if large-amplitude inhomogeneities already existed at the time of primordial nucleosynthesis ($t = 1-100$ s). This would not be expected according to most theoretical ideas, but is indeed possible in principle, as Professor McCrea proposes. Let me illustrate this by an (admittedly artificial) example. Suppose that the Universe were divided into 'cells', in half of which the baryon-to-photon ratio was 10 times higher than in the other half. If these were *isothermal* perturbations, they would involve only small-amplitude *metric* fluctuations even if the cell size corresponded to 10^3–$10^4 M_\odot$, because the mass–energy of the uniformly distributed radiation would overwhelmingly dominate that of the baryons at the stage when the cells were larger than the horizon. If all the high-density cells developed into supermassive stars which collapsed into black holes, then all the baryons we now observe would have acquired a chemical composition characteristic of a low density universe even though the actual mean baryonic density was *ca.* 10 times higher.

ome
Dependence of macrophysical phenomena on the values of the fundamental constants

By W. H. Press and A. P. Lightman

Harvard-Smithsonian Center for Astrophysics, Cambridge, Massachusetts 02138, U.S.A.

Using simple arguments, we consider how the fundamental constants determine the scales of various macroscopic phenomena, including the properties of solid matter; the distinction between rocks, asteroids, planets, and stars; the conditions on habitable planets; the length of the day and year; and the size and athletic ability of human beings. Most of the results, where testable, are accurate to within a couple of orders of magnitude.

1. Introduction

We have been given the pleasant task of reviewing the manner in which the fundamental constants affect our daily lives. Why are 'things' the way they are instead of some very different way? What sets the scale of the size of planets, the size of people, how fast we run, the length of the year, and so on?

This general area of diversionary physics has attracted the occasional attention of researchers for quite a while. Galileo considered scaling limits on living creatures, based on strength of materials arguments. More recently, we can trace a lineage including works by Haldane (1928), Weisskopf (1969, 1975), Salpeter (1966), Carter (1974), Carr & Rees (1979), and others. We will draw freely on this material in the present review.

The general organization of this paper will become clear to the reader as he or she proceeds. Roughly, the material is presented in order of increasing speculation.

2. Properties of solid material

(a) Atomic properties

Solid material has a dull and straightforward existence. It occupies a régime where only two characteristic masses are important, the mass of the proton m_p and the mass of the electron m_e. The only two forces of any importance are the electrostatic force of attraction, which brings in the electronic charge e and the degeneracy force or 'Pauli exclusion-principle force', which brings in Planck's constant $\hbar$. Relativistic effects are not important in this régime, so the speed of light c plays no role. Likewise, gravitational effects are not important, so Newton's constant G plays no role.

The basic scale of atomic distances is the Bohr radius a_0, where electric and degeneracy forces balance,

$$a_0 = \hbar^2/m_e e^2 \approx 0.53 \times 10^{-8} \text{ cm}, \tag{1}$$

and the basic scale of the energy in atomic processes is the binding energy of two charges separated by a Bohr radius, the rydberg

$$Ry = e^2/2a_0 \approx 2.2 \times 10^{-11} \text{ erg}\dagger = e^4 m_e/\hbar^2 = \alpha^2 m_e c^2. \tag{2}$$

† 1 erg = 10^{-7} J.

The final equality follows from the definition of the electromagnetic fine structure constant $\alpha = e^2/\hbar c \approx 1/137$, and shows why atomic processes are non-relativistic.

The characteristic frequency of atomic electronic transitions is

$$\omega_0 = Ry/\hbar \sim e^4 m_e/\hbar^3 \approx 4 \times 10^{16} \text{ s}^{-1}.\dagger \tag{3}$$

Equivalently, the wavelength of light emitted in atomic processes, $2\pi c/\omega_0$, is about $2\pi \times 137$ Bohr radii, which is somewhat shorter than 0.1 μm and is what we call ultraviolet.

The characteristic density of all solid matter is of the order of one proton mass per cubic Bohr diameter,

$$\rho_0 = m_p/(2a_0)^3 \approx 1.4 \text{ g cm}^{-3}. \tag{4}$$

The characteristic bulk modulus (compressibility) of all solid matter is of the order of its internal binding energy density, namely

$$\mu_b \sim Ry/(2a_0)^3 \approx 1.8 \times 10^{13} \text{ dyn cm}^{-2}\ddagger. \tag{5}$$

The characteristic coefficient of thermal expansion is

$$k/Ry \approx 6.3 \times 10^{-6} \text{ K}^{-1}. \tag{6}$$

Here k is Boltzman's constant, which we can usually avoid by writing all temperatures in units of energy (e.g. 1 eV $\approx 10^4$ K).

In order of magnitude, the opacity of matter to electromagnetic radiation is set by the Thomson cross section

$$\sigma_T = \tfrac{8}{3}\pi \, (e^2/m_e c^2)^2. \tag{7}$$

Measured as a total cross section per bulk mass, the opacity is

$$\kappa_T = \sigma_T/m_p \approx 0.4 \text{ cm}^2 \text{ g}^{-1}. \tag{8}$$

While opacities near to atomic resonances may be much larger than this, and much smaller far away from resonances when all electrons are tightly bound, equation (8) is generally the only dimensional scale that can enter.

(b) Molecular properties

The characteristic vibrational frequencies of many molecular bonds are smaller than those of electronic transitions by a factor $(m_e/m_p)^{\frac{1}{2}}$, basically because the entire mass of the molecule, not just the electronic cloud, participates in such modes. Correspondingly, the characteristic bond energies are down by the same factor from the rydberg. In typical solid crud (we use 'crud' in its technical sense, meaning a substance which is neither metallic nor single-crystalline), these weaker bonds are responsible for shear rigidity, and for holding the crud together under forces of extension. Therefore, the typical shear modulus or tensile strength of crud is set in scale by

$$\mu_s \sim [Ry/(2a_0)^3] \, (m_e/m_p)^{\frac{1}{2}} \approx 4 \times 10^{11} \text{ dyn cm}^{-2}. \tag{9}$$

In practical units, this is about 6×10^6 lb/in². § Real materials always contain structural flaws and are an additional factor of 10 or 100 below this.

---

† We use the symbol $\sim$ to mean 'of order'.
‡ 1 dyn = 10^{-5} N.
§ 1 lb $\approx$ 0.453 kg, 1 in = 2.54×10^{-2} m.

The characteristic thermal conductivity of solid crud is set by its lattice spacing and lattice vibrational frequency. As for all transport phenomena, the transport coefficient is the product of a number density of carriers (here of order a_0^{-3}), a mean free path or phonon wavelength (here of order a_0), a transport velocity (a_0 times equation (3) times $(m_e/m_p)^{\frac{1}{2}}$ divided by $\hbar$), and a specific heat per molecule k. The result is

$$\text{(conductivity)} \sim (Ry/a_0\hbar)\,(m_e/m_p)^{\frac{1}{2}}k \approx 1 \times 10^7 \text{ erg cm}^{-1}\text{ K}^{-1}\text{ s}^{-1}. \tag{10}$$

In fact, like (9), this estimate is about a factor of a hundred too large, since real materials are full of dislocations and other phonon-scattering unpleasantness.

For what follows below, we will need to take cognizance of the fact that an important set of complex chemical phenomena take place with bond energies that are even another factor of ten smaller than the bond energies in crud. This extra factor of about 0.1, which we will denote by the symbol ϵ, does not arise out of any combination of physical constants, but comes from all the abhorrent details of chemistry that are omitted in this paper. Substances that are made of these delicate bonds are called 'organic'. The characteristic bond energy of organic substances is then of order

$$\epsilon Ry(m_e/m_p)^{\frac{1}{2}}. \tag{11}$$

The temperature corresponding to this bond energy, i.e. such that kT is of order expression (11), is about 350 K. More on this in §4 below.

3. Self-gravitating objects

(a) Energy scalings

Gravity is so weak a force as to be negligible except when very large numbers of atoms are aggregated together. Let us consider an aggregation of N atoms in a configuration such that their mean separation is some distance d. Then the aggregate mass and size of the resulting object are

$$M \sim Nm_p \qquad R \sim N^{\frac{1}{3}}d. \tag{12}$$

The gravitational binding energy (g.e.) of the object is

$$\text{g.e.} \sim GM^2/R \sim G(Nm_p)^2/N^{\frac{1}{3}}d. \tag{13}$$

The degeneracy energy (d.e.) of the object is obtained by application of the uncertainty principal to get a Fermi momentum ($p_F d \sim \hbar$), and converting this to a Fermi energy by $E_F \sim p_e^2/(2m_e)$, giving

$$\text{d.e.} \sim N(\hbar/d)^2/2m_e. \tag{14}$$

If the object is at a finite temperature T, then it has a total thermal energy (t.e.) content of

$$\text{t.e.} \sim NkT, \tag{15}$$

and a total radiation energy (r.e.) given by the product of its volume with aT^4, where a is the radiation constant:

$$\text{r.e.} = Nd^3aT^4, \quad a \equiv \tfrac{1}{15}\pi^2\,k^4/c^3\hbar^3. \tag{16}$$

The object's electrons, which are the lightest particles and thus the first to become relativistic as temperature or density is increased, have a total rest-mass energy (e.m.e.)

$$\text{e.m.e.} \sim Nm_e c^2. \tag{17}$$

Finally, the total rest mass energy of the object is

$$\text{p.m.e.} \sim Mc^2 \sim Nm_\text{p}c^2. \tag{18}$$

We give the various classes of objects in the Universe different names according to which of the above forms of their energy dominate, which balance, which are negligible compared with the others, and which (if any) are comparable with the chemical or molecular binding energy densities of (5) or (9). We use such names as rocks, asteroids, planets, stars, black holes, and so forth.

(b) Rocks, asteroids, planets

Consider the classes of objects for which the degeneracy energy equation (14) dominates *all* the others when d has the value of the Bohr radius a_0. Then the degeneracy energy gives the object a bulk incompressibility given numerically by (5). As we saw in §2, such objects will consist of ordinary matter ('crud'), at ordinary density given by (4). We call such objects 'rocks' or, as they get larger, 'asteroids'. Gravitation is entirely unimportant for these objects.

There is no minimum size to a rock. There is, however, a maximum size to an asteroid, defined to be the size where gravitational forces are able to overcome the shear modulus of the material and thus cause it to become spherical instead of arbitrarily rock-shaped. We can calculate this size in order of magnitude by equating the gravitational energy (13) to the volume times the shear modulus (9), and using the relations (12). This gives a mass of

$$M \sim \tfrac{1}{64}(e^2/Gm_\text{p}^2)^{\frac{3}{2}} (m_\text{e}/m_\text{p})^{\frac{3}{4}} m_\text{p} \approx 1 \times 10^{26} \text{ g}. \tag{19}$$

This mass is the maximum mass of an asteroid or, equivalently, the minimum mass of a planet, a planet being defined as an object made of crud that is rendered spherical by gravitational forces. Equation (19) gives a result that is a rather larger than the actual number, because the actual shear strength of planetary matter is less than its dimensional value; that is geology, not physics, but we will need to say more about it below.

It is convenient to define the gravitational fine structure constant $\alpha_\text{G} \equiv Gm_\text{p}^2/\hbar c \approx 5.88 \times 10^{-39}$, so that one can write the first dimensionless combination on the right side of (19) as $(\alpha/\alpha_\text{G})^{\frac{3}{2}}$. Then, rewriting (19), the minimum mass and radius of a planet are, roughly,

$$M \sim m_\text{p}(\alpha/\alpha_\text{G})^{\frac{3}{2}} (m_\text{e}/m_\text{p})^{\frac{3}{4}}, \quad R \sim a_0(\alpha/\alpha_\text{G})^{\frac{1}{2}} (m_\text{e}/m_\text{p})^{\frac{1}{4}}. \tag{20}$$

The maximum mass of objects that we call planets is set by the condition that the gravitational energy does not dominate the degeneracy energy. When mass is increased so that the gravitational energy *does* dominate the degeneracy energy at normal density, then the object is called a 'cold, degenerate-dwarf star'. The dividing mass is given by equating (13) and (14), with $d = a_0$. This gives

$$M \sim (\alpha/\alpha_\text{G})^{\frac{3}{2}} m_\text{p} \approx 2 \times 10^{30} \text{ g}. \tag{21}$$

Comparing with (20), we see that planets exist for a mass range of only about $(m_\text{e}/m_\text{p})^{-\frac{3}{4}} \approx 300$. The low end of this range is somewhat below the mass of the Earth, while the high end of the range is at about the mass of Jupiter. (Jupiter is almost a stillborn star.)

(c) Cold, degenerate stars

Cold, degenerate stars range upwards in mass from the value given by (21), corresponding to d decreasing below a_0 and densities increasing above those of normal matter. We know very little about how many such objects there are in the Universe. At the low end of their mass range these objects will never have ignited their nuclear fuel, so they are dark and virtually unobservable. They may be the 'missing mass' that is observed gravitationally in galaxy rotation curves; that subject is one of current debate in astronomy.

There is an upper limit to the possible mass of cold, degenerate stars, first found by Chandrasekhar (1931a, b). The relation between mass and radius for such stars is obtained by equating (13) (gravitational binding energy) and (14) (degeneracy energy). One finds that the radius varies as the $-\frac{1}{3}$ power of mass. Along this sequence, the other forms of energy, equations (15)–(17), are negligible (in part by definition, since $T \equiv 0$). As the mass approaches some maximum value M_C, however, the degeneracy energy approaches the rest energy of the electrons, equation (17). When the electrons become relativistic, it can be shown (Landau 1932) that the solution of the equation of hydrostatic equilibrium gives an unstable, rather than a stable, solution.

This maximum mass M_C of cold, degenerate stars is thus obtained by equating (13), (14), and (17), eliminating d and solving for N. The result is the *Chandrasekhar mass*

$$M_C \sim (\hbar c/Gm_p^2)^{\frac{3}{2}} m_p = \alpha_G^{-\frac{3}{2}} m_p \approx 3.7 \times 10^{33} \text{ g}. \tag{22}$$

Notice that the Chandrasekhar mass is a factor of $(137)^{\frac{3}{2}} \approx 1600$ larger than (21), so this is the mass range over which cold, degenerate stars are possible.

As we will now see, aggregations of matter at the upper end of this range *are* capable of igniting their nuclear fuel. In the real Universe, therefore, cold, degenerate dwarfs in the upper part of the range occur as the cinders of formerly luminous stars and not as primordial objects.

(d) Luminous stars

A luminous star, as opposed to a cold degenerate star, is one in which thermal energy dominates degeneracy energy. Gravitational force is balanced with thermal pressure, so the gravitational energy (13) and thermal energy (15) must be about equal. That gives one relation between the mass M, radius R, and average (or central) temperature T of the star. We need one further relation before we can solve for the radius and temperature of stars as a function of their mass.

The extra relation is an equation that relates the central temperature of a star to the *logarithm* of fundamental constants, R, and M. The logarithm is so slowly varying that it is, for all order-of-magnitude purposes, a constant. (One can solve for the self-consistent values of R and M that go into the logarithm by iteration, starting with virtually any first guesses.)

The key idea is this: we call something a star if the time that it takes to burn all its nuclear fuel is longer than its thermal diffusion time. This is the condition for it to be able to settle down to a radiative equilibrium. Something that releases nuclear energy faster than its thermal diffusion time is a nuclear explosion, not a star.

The thermal diffusion time is the time that it takes a photon to random-walk its way through the distance R. This equals the light travel time across that distance times the optical depth. The optical depth is related to M, R, and the opacity (8) by

$$\tau_{\text{opt}} \sim M\sigma_T/R^2 m_p, \tag{23}$$

so the radiative diffusion time (also called the Kelvin–Helmholtz time) is

$$t_{\rm KH} \sim M\sigma_{\rm T}/Rcm_{\rm p} \approx 1 \times 10^4 \text{ a}, \tag{24}$$

where the numerical value is calculated using $M = 2 \times 10^{33}$ g, $R = 7 \times 10^{10}$ cm, the actual values for the Sun. The above timescale is actually about 100 times too small, due to our underestimate of the opacity by that factor.

The rate of nuclear burning, that is the inverse lifetime of a typical particle against a nuclear reaction, is given by the familiar expression involving density, particle velocity, and cross section $\sigma_{\rm nuc}$

$$\lambda_{\rm nuc} \sim (M/R^3)\,\sigma_{\rm nuc} v/m_{\rm p}, \tag{25}$$

where v is a typical thermal velocity of order $(kT/m_{\rm p})^{\frac{1}{2}}$. (We should note here that detailed stellar models are centrally condensed and have central densities that are about 100 times larger than M/R^3.)

The nuclear cross section is that appropriate for weak interactions with a Coulomb barrier (e.g. $p + p \rightarrow p + n + e^+ + \nu$). Some detailed nuclear kinematics shows that this varies with energy E as (see, for example, Clayton 1968)

$$\sigma_{\rm nuc} = (S/E) \exp(-2\pi e^2/\hbar v), \tag{26}$$

where S is the so-called 'astrophysical cross section factor'. The calculation of S from first principles will bring in the weak interaction constant $G_{\rm F} \sim \alpha/m_{\rm W}^2$, where $m_{\rm W}$ is the mass of the intermediate boson, as well as various phase space factors involving $m_{\rm p}$ and $m_{\rm e}$. Since, however, we will only need the logarithm of S, we will not try to track through these details, but rather use its measured value, which is on the order of 10^{-21} keV b.†

We now get the central temperature of a luminous star by setting the product of (25) and (24) equal to unity and then taking the logarithm. This gives, collecting factors together,

$$\frac{2\pi e^2}{\hbar (kT/m_{\rm p})^{\frac{1}{2}}} \sim \ln\left[10^8 \times \left(\frac{M}{m_{\rm p}}\right)^2 \left(\frac{\sigma_{\rm T}}{R^2}\right) \left(\frac{kT}{m_{\rm p}c^2}\right)^{\frac{1}{2}} \left(\frac{S}{R^2 kT}\right)\right]. \tag{27}$$

The factor 10^8 has been inserted into the logarithm to compensate for the underestimates of $t_{\rm KH}$ and $\lambda_{\rm nuc}$ which were already mentioned, and also to reflect the fact that most objects that we call stars live much *longer* than one Kelvin–Helmholtz time. In part, this is observational selection: we see the relatively long-lived objects. For 'correct' values of R, M, and T, and S, the logarithm has the value of about 15, which can be found self-consistently once M and R are determined.

Rewriting (27), we obtain the central temperature of all stars,

$$kT_* \approx m_{\rm p}c^2(2\pi\alpha^2)\,(\tfrac{1}{15})^2 \sim 0.1\alpha^2 m_{\rm p}c^2 \approx 10 \text{ keV}. \tag{28}$$

Actual central temperatures are about a factor 10 smaller than this, due in part to the fact that most reactions occur several e-folds out on the tail of the Maxwellian thermal distribution. The above derivation is essentially that of Salpeter (1966). Wherever the explicit factor 0.1 appears below, it is to be traced back to the logarithmic numerical factor in (28). We now can equate (13) and (15) and obtain the linear relation between the mass and the radius of luminous (main-sequence) stars,

$$M/R \sim 0.1c^2\alpha^2/G \approx 7 \times 10^{22} \text{ g cm}^{-1}. \tag{29}$$

† 1 b = 10^{-28} m².

Since radiation density, which is the conserved quantity diffused through the star, goes as T^4, the surface temperature T_s of a star of mass M is down from its central temperature by a factor of $\tau_{\text{opt}}^{\frac{1}{4}}$, where τ_{opt} is given by (23). So,

$$T_s \sim \frac{T_*}{\tau_{\text{opt}}^{\frac{1}{4}}} \sim 0.1 \frac{m_p c^2}{k} \alpha^2 \left(\frac{R^2 m_p}{M^2 \sigma_T}\right)^{\frac{1}{4}} M^{\frac{1}{4}}$$

$$\sim \sqrt{(0.1)} \alpha \frac{m_p c^2}{k} \left(\frac{\alpha_G}{\alpha}\right)^{\frac{1}{2}} \left(\frac{m_e}{m_p}\right)^{\frac{1}{2}} \left(\frac{M}{m_p}\right)^{\frac{1}{4}} \approx 10^5 \text{ K}, \quad (30)$$

where (29) has been used in going from the first line to the second. Actual stars have surface temperatures about a factor ten less than this estimate, primarily due to more complicated effects in the opacity.

The luminosity of a star of mass M follows from (30), (29), and from the radiation law,

$$L \sim ac T_s^4 R^2 \sim (c^5/G) \alpha_G^5 \alpha^{-2} (m_e/m_p)^2 (M/m_p)^3. \quad (31)$$

Equation (31) is an unlucky case where a slightly more detailed treatment (Lightman 1983) shows that neglected factors of order unity have built up to a significant factor. The factor $\pi/(3^3 \times 2^5 \times 15) \approx 2 \times 10^{-4}$ should be inserted in front of the right side. With this factor, (31) gives numerically (for the Sun's mass) about 6×10^{33} erg s^{-1}, as compared with the actual value of about 4×10^{33} erg s^{-1}. It is interesting to note that, although the equations which led to (31) contained the assumed central temperature (28), that temperature actually cancels out of (31).

So we now know how the properties of stars vary with their mass. But what are the upper and lower limits to the masses of stars along this natural sequence?

The lower limit is set by the requirement that a luminous star that satisfies the mass–radius relation (29) have a degeneracy energy (14) that is smaller than its gravitational or thermal energy content. If this were not so, then the star would never have condensed down to the point of igniting its nuclear fuel: it would have come to equilibrium as a cold, degenerate star at a larger radius than that required by (29). Manipulating (14), (12), and (29), we can write this minimum mass as

$$M \sim (0.1 m_p/m_e)^{\frac{3}{4}} (\alpha/\alpha_G)^{\frac{3}{2}} m_p \approx 1 \times 10^{32} \text{ g}, \quad (32)$$

which is larger than the maximum mass of a planet or minimum mass of a cold, degenerate star (21) by the factor $(0.1 \times 1836)^{\frac{3}{4}} \approx 50$.

The maximum mass of a star is set by the requirement that the total energy of thermal radiation (16) should not dominate the thermal energy (15). If this becomes violated, then the star, now pressure-supported by a relativistic photon gas, becomes hydrostatically unstable, much as the degenerate stars did when their electrons became relativistic. The limiting mass is thus obtained by equating

$$(M/m_p) kT \sim GM^2/R \sim R^3 a T^4. \quad (33)$$

If T is eliminated between these equations, one finds miraculously that R also cancels out, giving a mass limit independent of our assumed central temperature,

$$M \sim \alpha_G^{-\frac{3}{2}} m_p \approx 4 \times 10^{33} \text{ g}. \quad (34)$$

This is dimensionally just the same as the Chandrasekhar mass limit for cold, degenerate stars, although the physical situation contemplated is quite different. Because of neglected factors of order unity in (34), one is actually able to find stars up to about ten times the indicated mass.

Comparing (34) with (32), we see that luminous stars are possible over a mass range of only about $137^{\frac{3}{2}}/50 \approx 30$, ranging down from somewhat larger than the Chandrasekhar mass.

4. Habitable planets

We are now in a position to say something not just about planets in general, but about planets with an environment conducive to the evolution of complicated organic structures ('organic' having been defined by (11)).

(a) Mass and radius of the Earth

Since the Earth is made of ordinary crud, its mass M and radius R lie on the ordinary density relation (4),

$$M/R^3 \sim m_p/(2a_0)^3. \tag{35}$$

We need an additional relation that distinguishes planets in general from habitable planets. That relation follows from positing that a habitable planet must have a differentiated atmosphere (neither vacuum nor primordial hydrogen and helium) that is gravitationally bound to the planet at the temperature of expression (11), where interesting organic reactions can take place. Therefore, the escape velocity from the surface of a habitable planet should be greater, but not too much greater, than the thermal velocity of hydrogen at that temperature (Press 1980). This condition gives

$$GM/R \sim \epsilon(m_e/m_p)^{\frac{1}{2}} Ry/m_p. \tag{36}$$

Now, (36) and (35) can be solved for M and R separately, giving

$$\left. \begin{array}{l} R \sim \epsilon^{\frac{1}{2}}(2a_0)(m_e/m_p)^{\frac{1}{4}}(\alpha/\alpha_G)^{\frac{1}{2}} \approx 5 \times 10^8 \text{ cm}, \\ M \sim \epsilon^{\frac{3}{2}}m_p(m_e/m_p)^{\frac{3}{4}}(\alpha/\alpha_G)^{\frac{3}{2}} \approx 2.5 \times 10^{26} \text{ g}, \end{array} \right\} \tag{37}$$

where the numerical values are evaluated using $\epsilon = 0.1$ (see discussion preceeding expression (11)). These values are in moderately good agreement with the true values for the Earth, $R \approx 6.4 \times 10^8$ cm and $M \approx 5.9 \times 10^{27}$ g. Notice that the mass of a habitable planet is smaller than the maximum planetary mass (equation (21)) by a factor $\epsilon^{\frac{3}{2}}(m_e/m_p)^{\frac{3}{4}}$.

(b) Length of the day and of the year

All planets in the solar system, unless they are tidally locked to some other planet (Venus), a nearby moon (Pluto), or the Sun (Mercury), are observed to rotate at an angular velocity that is within an order of magnitude of centrifugal breakup. While the evolutionary details that lead to this circumstance are not well understood, the phenomenon seems likely to be universal. There is, in fact, an abundance of angular momentum in observed protostellar gas clouds. The angular velocity of centrifugal breakup depends only on the density, not on the mass and radius separately, and is of order $G\rho^{\frac{1}{2}}$. The length of a 'universal day' therefore follows from equation (4) (Lightman 1983),

$$T_{\text{day}} \sim \frac{2\pi}{G\rho_0^{\frac{1}{2}}} \sim 2\pi(2)^{\frac{3}{2}} \frac{a_0}{c} \left(\frac{m_p}{m_e}\right)^{\frac{1}{2}} \alpha^{-\frac{1}{2}} \alpha_G^{-\frac{1}{2}} \approx 6 \text{ h}. \tag{38}$$

The length of the year is not, of course, universal for all planets. However, its value is determined for habitable planets, as we have defined them: from (31), we know the luminosity of a star as a function of its mass. Equations (32) and (34) bracket the possible mass range of luminous stars into a narrow range. Let us take the geometric mean of these expressions as being characteristic of the 'typical' star, $M \sim 0.1 m_p \alpha_G^{-\frac{3}{2}}$. Now we can compute the distance d

of a habitable planet from its star by the requirement that it be in thermal equilibrium at the habitable temperature of (11). This, together with the black body law, and the above typical mass of a star, gives a determination of the 'astronomical unit' d in terms of fundamental constants

$$d \sim \left(\frac{60}{\pi^3 10^3}\right)^{\frac{1}{2}} a_0 \left(\frac{m_p}{m_e}\right) \epsilon^{-2} \alpha^{-4} \alpha_G^{-\frac{1}{4}} \approx 5 \times 10^{13} \text{ cm}. \tag{39}$$

When the above expressions for M and d are combined with Kepler's law, we obtain an expression for the year on a habitable planet in terms of fundamental constants (Lightman 1983)

$$T_{\text{year}} \sim 0.2(a_0/c) \, (m_p/m_e)^2 \, \epsilon^{-3} \alpha^{-\frac{13}{2}} \alpha_G^{-\frac{1}{8}} \approx 20 \text{ a}. \tag{40}$$

All biological phenomena associated with the changing seasons may thus be expected to follow the universal time scale given by (40).

(c) Height of mountains on earth

Weisskopf (1975) has treated this topic beautifully: the pressure at the base of a mountain of height H is the product of its height, its density, and the acceleration of gravity. We equate this to the shear strength, (9), giving

$$H(GM/R^2) \rho_0 \sim [Ry/(2a_0)^3] \, (m_e/m_p)^{\frac{1}{2}}. \tag{41}$$

With equation (4) for universal density, both for the mountain and for the planet, (41) can be rewritten as the ratio of the mountain height to the planetary radius,

$$H/R \sim (\alpha/\alpha_G) \, (m_e/m_p) \, (m_p/M)^{\frac{2}{3}}. \tag{42}$$

Now, something strange turns out if we substitute into (42) expression (37) for the mass of a habitable planet. We get an answer greater than one! Mountains can apparently be larger than the Earth! This is a case where we have once again run afoul of a build-up of dimensionless numbers that do not involve the fundamental constants. In fact, observationally, (37) is about right, while (42) is a big overestimate. Weisskopf does a more careful job than we are doing here, and shows that a factor $A^{-\frac{5}{3}}$ appears on the right of (42), where A is the mean molecular weight of the material (around 50 for both Fe and SiO_2); Weisskopf also tracks through additional materials-related factors amounting to an additional factor of about 0.03. With these inserted, (42) is not too bad.

In fact, we could have pointed out the problem when we first derived (37) above, since that mass of a habitable planet is smaller than the minimum mass of a planet (20), by a factor $\epsilon^{\frac{3}{2}}$. Equation (20) is an overestimate. On geological timescales, matter is softer than the fundamental constants indicate at first glance, and by a significant factor. (One can imagine a more thorough treatment, where the rate of plastic flow of rock at habitable temperature is derived from fundamental constants, and is taken to be in equilibrium with the age of the earth. On *very* long time scales, as Dyson (1979) has pointed out, all solid objects become spherical.)

(d) Sunlight and weather

We have already seen that the solar flux at Earth is determined by the condition that the Earth be in approximate black-body equilibrium at habitable temperatures. There is also something else remarkable about sunlight: its spectral temperature (that is, the surface temperature of the Sun) is smaller than, but on the order of, a rydberg. Life would be very difficult were this not the case. If the spectral temperature of sunlight were much higher, then organic structures could not withstand its flux; it would effectively sterilize our planet. If the spectral temperature were much lower, then sunlight could not easily drive photosynthetic reactions (see, for example, Wald 1959).

In part, the happy spectral coincidence is not a coincidence at all, but is due to details of the opacity of solar material which we have neglected. Opacities get large at a fraction of a rydberg, *because* there is a lot of chemistry there, and we see into the Sun just to the point where the opacities get large, by definition of optical depth. However, if stellar surface temperatures estimated by (30) were very much larger, then the opacity would not have the opportunity to get large anywhere near the stellar surface. A numerical coincidence of the fundamental constants is involved: dividing (30) by (2) gives, after some simplification,

$$kT_s/Ry \sim \sqrt{(0.1)} \, (m_p/m_e)^{\frac{1}{2}} \alpha_G^{\frac{1}{8}} \alpha^{-\frac{3}{2}} \approx 0.3. \tag{43}$$

There is presumably no fundamental reason that this combination of constants had to come out to be of order unity.

So now the Sun is shining, and we can turn to the weather. The temperature of the atmosphere, as we saw, is by definition the habitable temperature $\epsilon(m_e/m_p)^{\frac{1}{2}} Ry/k$. What is the pressure of the atmosphere at the Earth's surface? We do not know of a believable estimate based on fundamental considerations. Comparing the Earth with Venus and Mars, whose atmospheric pressures are respectively of order $10^{\pm 2}$ times that of Earth, we see that the range of pressure can be large. It may be that the only answer is an anthropic one: our atmospheric column density is not too different from the reciprocal of (8), i.e. that density which provides a significant, but not overwhelming, opacity. Some opacity, we might argue, may be necessary to screen ultraviolet and cosmic radiation, while too much would produce a surface environment with no source of spectral free energy, just a local thermodynamic equilibrium at the habitable temperature. We would not wish to lean too heavily on these arguments.

It is somewhat easier to see what sets the scale of velocity of large-scale planetary winds on the Earth. The speed of sound in air follows from the habitable temperature,

$$v_s \sim (kT/m_p)^{\frac{1}{2}} \sim \epsilon^{\frac{1}{2}}\alpha(m_e/m_p)^{\frac{3}{4}} c \approx 2 \times 10^5 \text{ cm s}^{-1}. \tag{44}$$

It is not a coincidence that this value is about the same as the Earth's rotational velocity. That just restates the facts that the Earth rotates not too far from breakup velocity (equation (38)), and that the atmosphere is not too gravitationally bound at habitable temperature (equation (36)). By simple geometry, the black-body equilibrium temperature at extreme latitudes should be smaller than that at the equator by of order $2^{\frac{1}{4}}$ (the fourth-root coming from the usual radiation law). This differential, if not compensated by wind motion, would generate pressure differences of the same fractional amount, and cause pressure gradient driven winds on the order of a like fraction of the speed of sound. But since the rotational speed of the Earth is also of order the speed of sound, the geostrophic (circular) flow which balances those pressure gradients has comparable speed, and has a scale on the order of the size of the planet. So, winds must be of order some modest fraction of the sound speed. As we know from observation, that fraction is on the order of a few tenths (on planetary scales), scaling down by some Kolmogorov-like (but not fully three-dimensional) cascade on smaller scales. So, the dependence of wind velocity on the fundamental constants is the same as in (44).

5. The human condition in terms of fundamental constants

(a) How big are we?

While we might wish to define ourselves as humans according to our ability to reason, the physics of the matter is more prosaic: we are as highly evolved as we are because we are about as big as we can be without breaking when we fall. This criterion may not distinguish us from horses and elephants; as far as the fundamental constants know, they are human too. We can compute the implied human size h and mass $m \sim h^3 \rho_0$ by equating our potential energy in a gravitational field of acceleration g to our breaking energy, the total bond energy of a 2-dimensional surface containing $(m/m_p)^{\frac{2}{3}}$ organic bonds (Press 1980):

$$mgh \sim \epsilon (m_e/m_p)^{\frac{1}{2}} Ry (m/m_p)^{\frac{2}{3}}. \qquad (45)$$

Substituting the mass and radius of the Earth (equation (37)) into g gives, after some rearrangement,

$$h \sim \epsilon^{\frac{1}{4}} (m_e/m_p)^{\frac{1}{4}} (2a_0) (\alpha/\alpha_G)^{\frac{1}{4}} \approx 3 \text{ cm}. \qquad (46)$$

This number is observationally a factor of about 10^2 too small. The reason is that, rather as mountains were considerably softer than their dimensional strength, so we are considerably tougher. We are composed of a polymeric molecular structure that distributes shocks over a larger region than the weakest fault-plane of brittle, semi-crystalline crud.

Neglecting the smaller numerical factors, one sees that the human size (46) lies at the geometric mean of the planetary size (equations (20) or (37)) and the atomic size a_0 (Carr & Rees 1979). Also noteworthy is that the right side of (46) is also obtained when one computes either (i) the size of first-destabilized water waves on a lake (Weisskopf 1975), or (ii) the maximum size of water drops dripping off a ceiling. Both of those calculations involve a somewhat different physical situation, equation of surface tension to gravitational stress, but the dimensional combinations that result are the same. In a loose sense, this is why there is something special about the centimetre–metre scale on Earth, why the range of phenomenology is so rich on that scale: both solid and liquid 'things' have material strengths on about that scale.

(b) How hard can we work?

As already mentioned, there is no fundamental distinction between humans and horses. It is of interest, therefore, to compute the horsepower in terms of fundamental quantities.

Over a very large range of animal sizes, the peak power output of animals is observed to scale as a power not too different from the $\frac{2}{3}$ power of their mass (Wilkie 1959). Likewise, the resting metabolism of animals scales with mass with the same exponent. These data strongly suggest that the power output is limited by cooling through the animal's surface area, and that resting metabolism is scaled to keep the resting animal sufficiently warm.

In this case, we can compute the horsepower in terms of a tolerable thermal differential ΔT,

$$\text{(horsepower)} \sim \Delta T \times \text{(conductivity)} \times \text{(area)}/\text{(skin depth)}. \qquad (47)$$

If we put in observed values 10 °C, 4×10^4 erg °C^{-1} cm^{-1} s^{-1}, 1 m^2, and 1 cm, (47) gives about 400 W, which is close to the observed value.

If we had no knowledge of the observed parameters, we could use $\Delta T \approx T$, area of order h^2, skin depth of order h (where h is given by equation (46)), and conductivity as given by (10). These values are

$$\text{(horsepower)} \sim e^{\frac{5}{4}}(Ry)^2/\hbar\, (m_e/m_p)^{\frac{3}{4}}\, (\alpha/\alpha_G)^{\frac{1}{4}} \approx 200\,\text{W}. \tag{48}$$

By equating skin thickness to h, rather than to a constant for all animals, we have unfortunately lost the proper scaling law (horsepower) $\propto h^2 \propto m^{\frac{2}{3}}$.

(c) How fast can we run?

We move in an extremely dissipative fashion. To run at a velocity v, we must renew practically our whole kinetic energy every stride, that is, every motion through our own body length h. The power needed to run at velocity v is therefore of order mv^3/h. Equating this to the available power of (48), and using (46), we can write the velocity of a human runner as dimensionless factors times the speed of light.

$$v \sim e^{\frac{1}{4}}(\alpha_G/\alpha)^{\frac{1}{12}}\, \alpha(m_e/m_p)^{\frac{7}{12}}\, c \approx 15\,\text{m s}^{-1}: \tag{49}$$

not too bad an estimate, which shows that the four minute mile has a more fundamental significance than is commonly supposed.

This work was supported in part by NSF grant PHY-80-07351, and by NASA grant NAGW-246.

References

Carr, B. J. & Rees, M. J. 1979 *Nature, Lond.* **278**, 605.
Carter, B. 1974 In *Confrontation of cosmological theories with observation* (ed. M. S. Longair). Dordrecht: Reidel.
Chandrasekhar, S. 1931a *Phil. Mag.* **11**, 592.
Chandrasekhar, S. 1931b *Astrophys. J.* **74**, 81.
Clayton, D. D. 1968 *Principles of stellar evolution and nucleosynthesis*, ch. 4 and 5. New York: McGraw-Hill.
Dyson, F. J. 1979 *Rev. mod. Phys.* **51**, 447.
Haldane, J. B. S. 1928 In *Possible worlds*. New York: Harper. (Reprinted in *The world of mathematics* (ed. James R. Newman) 1956 New York: Simon and Schuster.
Landau, L. D. 1932 *Phys. Z. SowjUn.* **1**, 285.
Lightman, A. P. 1983 *Am. J. Phys.* (In the press.)
Press, W. H. 1980 *Am. J. Phys.* **48**, 597.
Salpeter, E. E. 1966 In *Perspectives in modern physics* (ed. R. E. Marshak). New York: John Wiley and Sons.
Wald, G. 1959 *Scient. Am.* **201**, 92.
Weisskopf, V. F. 1969 Lectures given in the CERN summer vacation program. CERN preprint 70–8. Geneva.
Weisskopf, V. F. 1975 *Science, N.Y.* **187**, 605.
Wilkie, D. R. 1959 *Nature, Lond.* **183**, 1515.

Discussion

Sir Rudolf Peierls, F.R.S. (*Nuclear Physics Laboratory, Keble Road, Oxford OX1 3RH, U.K.*). Does not the estimate of molecular binding energies used really represent an estimate of the zero-point energy of vibration? If so, it would seem a considerable underestimate, since the binding energy usually amounts to many vibrational quanta.

W. H. Press. Yes, the factor of $(m_e/m_p)^{\frac{1}{2}}$ does strictly give the characteristic energy-level *spacing* of molecules, rather than their full binding energy. Numerically, however, it does also give the correct (rough) factor by which molecular bindings are smaller than typical atomic ones. One may wish to consider the factor a *mnemonic* for this dimensionless ratio of binding energies (which derives from the details of chemistry) rather than as an accurate physical 'theory' of that ratio.

T. GOLD, F.R.S. (*Space Sciences Building, Cornell University, Ithaca, New York* 14853, *U.S.A.*).
This is to discuss some of the implications of suggestions that 'constants' of nature change with time (suggestions that do not seem very probable to me). There are two distinct possibilities to be discussed. First, there is the case in which local measurements can show the change. This means, of course, that non-dimensional numbers, constructed from the 'constants', will change with time. It is then possible to register everywhere an 'absolute time', meaning that everywhere each instant is defined by the value that one of the changing non-dimensional numbers has assumed.

Such an 'absolute time' may make some conflict with relativity theory; at least it would define a 'preferred' system of time running through each space-like domain, defining an ordering of events that would not have been defined in any other way.

How should we assume this information about the value of the constants of nature is distributed? One possibility would be that it is the matter in each locality that carries a clock to register the time elapsed since the big bang; and that the value of the natural constants displayed by that matter corresponds to that elapsed time. But, of course, that time would not be unique in any one location. Different components of the matter in that location would have reached their present positions by different paths, and therefore the elapsed time since the big bang that they would register would be different. We have never found any evidence for particles of the same type to have any properties that would make them non-identical, and much of physics has been successfully constructed on the presumption that they are identical. I suppose most physicists would reject the idea of particles possessing clocks that show different times.

The other possibility is that a field pervades space, that is created by the universe, and that is dependent in some way on its structure at each moment of time. It would be somewhat analogous to the inertia-field with which one may wish to satisfy Mach's considerations.

This would be a case where one thinks of the structure of the universe and the laws of physics being an interrelated system. In a big bang cosmology the possibilities for any such interrelation are very limited. The observations that the ratio of spectral frequencies seen in distant galaxies accords with the local values suggests that most of local physics there, and at the epoch concerned, is the same as here. This does not prove, however, that the gravitational fine structure constant has the same value there as here. I would consider it as rather artificial to have one natural constant dependent on the structure of the Universe, and the others independent of it; it must be admitted, however, that such a possibility cannot be ruled out with the considerations mentioned. It seems unlikely that energy and momentum would be locally conserved quantities in such a system.

The other possible case of the constants of nature changing, is the one in which no *local* measurement could reveal a change: where all non-dimensional constants would remain fixed. In this case we know that there is still the possibility of a change in the clock-rates in one locality, as compared with the clock-rates in another. This is the case known from the gravitational 'Einstein shift', but one could contemplate that such a shift may occur for other reasons also. It is known that all local physics remains unchanged, even if the change has the character that an *acceleration* in the clock-rate in our locality would be noted, if observed from afar. An example of this is provided by a clock situated inside a spherical mass-shell. It is there in a field-free space, and all local physics is independent of the presence of this shell. A contraction or expansion of the shell will result in a change in the clock-rate, as observed from far away.

If any 'cosmological Einstein shift' existed, it would alter the spectral frequencies observed in the distance; so long as the observations span too short a time to note the cumulative change in distance that a relative velocity would create, any such shift would be indistinguishable from a doppler shift. In practice it would be a contribution of unknown magnitude to the observed redshift.

While such a non-gravitational shift would leave local dynamics unchanged, it is not clear what it would do to dynamics on a cosmological scale.

Dimensionality

By J. D. Barrow

Astronomy Centre, University of Sussex, Brighton BN1 9QH, U.K.

We examine the role played by the dimensions of space and space–time in determining the form of various physical laws and constants of Nature. Low dimensional manifolds are also seen to possess special mathematical properties. The concept of fractal dimension is introduced and we discuss the recent renaissance of Kaluza–Klein theories obtained by dimensional reduction from higher dimensional gravity or supergravity theories. A formulation of the anthropic principle is suggested.

1. Historical introduction

The fact that we perceive the world to have three spatial dimensions is something so familiar to our experience of it that we seldom pause to consider the influence this special property has upon the laws of physics. Yet, it appears that the dimensionality of the world plays a key part in determining the form of the laws of physics and in fashioning the roles played by the constants of Nature.

Interest in explaining why the world has three dimensions is by no means new. From the commentary of Simplicius and Eustratius (see Neugabauer 1975), Ptolemy is known to have written a study on the 3D nature of space entitled *On dimensionality* in which he argued that no more than three spatial dimensions are possible in Nature. Unfortunately this work has not survived. What does survive is evidence that the dramatic difference between systems identical in every respect but spatial dimension was discovered and appreciated by the early Greeks. The Platonic solids, first discovered by Theaitetos (Sarton 1959), brought them face-to-face with a dilemma: why are there an infinite number of regular, convex, two-dimensional *polygons* but only five regular three-dimensional *polyhedra*? This mysterious property of physical space later spawned many mystical and metaphysical 'interpretations'.

In the modern period, mathematicians did not become actively involved in attempting a rigorous formulation of the concept of dimension until the early nineteenth century. During the nineteenth century Möbius considered the problem of superimposing two enantiomorphic solids by a rotation through 4-space and later, Cayley, Riemann and others developed the systematic study of N-dimensional geometry although the notion of dimension they employed was entirely intuitive. It sufficed for them to regard dimension as the number of independent pieces of information required for a unique specification of a point in some coordinate system. Gradually the need for something more precise was impressed upon mathematicians by a series of counter-examples and pathologies to several simple intuitive notions. For example, Cantor and Peano provided injective and continuous mappings of $\mathbf{R}$ into $\mathbf{R}^2$ to confute ideas that the unit square contained more points than the unit line. After unsuccessful attempts by Poincaré it was Brouwer who established the key result (Brouwer 1911; 1913): he showed that there is no continuous injective mapping of $\mathbf{R}^N$ into $\mathbf{R}^M$ if $N \neq M$. The modern definition of dimension due to Menger and Urysoln grew out of this fundamental idea (Menger 1928; Hurewicz & Wallman 1941).

The question of the *physical* relevance of spatial dimension seems to arise first in the early work

(1747) of Immanuel Kant (see Handyside 1929). He realized that there was an intimate connection between the inverse square law of gravitation and the existence of precisely three spatial dimensions, although he regards the three space dimensions as a consequence of Newton's inverse square law rather than vice versa.

In the twentieth century a number of outstanding physicists have sought to accumulate evidence for the unique character of physics in three dimensions. Ehrenfest's famous article of 1917 was entitled *In what way does it become manifest in the fundamental laws of physics that space has three dimensions?* and it explained how the existence of stable planetary orbits, the stability of atoms and molecules, the unique properties of wave operators and axial vector quantities are all essential manifestations of the dimensionality of space. Soon afterwards Herman Weyl (1922) pointed out that only in $(3+1)$-dimensional space–times can Maxwell's theory be founded upon an invariant integral form of the action; so, only in $(3+1)$ dimensions is it conformally invariant.

In recent years the problem of dimensionality has re-emerged in connection with supersymmetric gauge theories and the description of chaotic dynamical systems. Our aim here is to give a brief discussion of some of the most interesting influences of dimensionality on physics and mathematics.

2. Orbits

One of the first areas of physics to display the role of dimensions is the theory of orbital motion under central forces. We immediately see the reason for the prevalence of inverse square laws in physics. Consider first the question of planetary motions.

The Poisson–Laplace equation for the gravitational field of force in an N-dimensional space has a solution for the gravitational potential, ϕ and force, F, of the form

$$\phi(r) \propto r^{2-N}; \quad F(r) \propto r^{1-N}, \quad N > 2, \tag{1}$$

for a radial distribution of material. The inverse square law of Newton follows as an immediate consequence of the tri-dimensionality. A planetary motion can only describe a central elliptic orbit in a space with $N = 3$ if its path is *circular*, but such a configuration is unstable to small perturbations. In three dimensions, of course, stable *elliptical* orbits are possible. If hundreds of millions of years in stable orbit around the Sun are necessary for planetary life to develop then such life might only develop in a three-dimensional world. In general, the existence of stable, periodic orbits requires $r^3 F(r) \to 0$ as $r \to 0$ and $r^3 F(r) \to \infty$ as $r \to \infty$. Therefore we require $N < 4$. If we examine the analogous problem in general relativity by examining motion in the gravitational field of the $(N+1)$ dimensional Schwarzschild solution we again find no stable bound orbits exist for $N > 3$.

One of Newton's most famous results is his proof that if two spheres attract each other under an inverse square law of force they may both be replaced by points concentrated at the centre of each sphere with mass equal to that of the respective sphere. It can be shown (Sneddon & Thornhill 1949) that the general form of the gravitational potential for which the gravitational force of a sphere can be replaced by that of a point at its centre is the Yukawa potential

$$\phi \propto e^{-\lambda r}/r, \tag{2}$$

although, for a sphere of radius a and density ρ, the mass that must be concentrated at its centre is

$$M(\lambda) = (4\pi a\rho/\lambda^2)(\cosh \lambda a - (\sinh \lambda a)/\lambda a), \tag{3}$$

and we recover Newton's result as $\lambda \to 0$

$$M(0) = \tfrac{4}{3}\pi a^3 \rho. \tag{4}$$

The truth of Newton's result is a direct consequence of the existence of three spatial dimensions in (1). Gravitation physics is simplest in three dimensions.

3. Atomic stability

It is widely known that matter is stable. By this we mean that the ground state energy of an atom is finite. However, the common text-book argument that employs the uncertainty principle to demonstrate this is actually false. Although the energy equation for a single electron of mass m and charge $-e$ in circular orbit around a charge $+e$ gives a total atomic energy of

$$E = h^2/2mr^2 - e^2/r, \tag{5}$$

and this energy apparently has a finite minimum of $r_0 \approx h^2/2me^2$, where $E'(r_0) = 0$, it is, in principle, possible for the electron to be distributed in a number of widely-separated wave packets. The packet close to the nucleus could then have an arbitrarily sharp momentum and position specification at the expense of huge uncertainty in the other packets. In this manner the ground state energy might be made arbitrarily negative.† For these reasons analyses of atomic stability such as those of Ehrenfest (1917) and those that use only the uncertainty principle must be regarded as only heuristic. However, their results are confirmed by an exact analysis of the Schrödinger equation. In 1917, Ehrenfest considered only the simple Bohr theory of an N-dimensional hydrogen atom. He found the energy and radii of the energy levels and noted that when $N > 5$ the energy levels increase with quantum number whereas the radii of the Bohr orbits

$$r_L(N) \approx (me^2 L^{-2} h^{-2})^{1/(N-4)}$$

decrease with increasing quantum number L and electrons just fall into the nucleus. Alternatively, if we write down the total energy for the system and use the uncertainty principle to estimate the kinetic energy resisting localization we have for $N > 2$, that

$$E \approx h^2/2mr^2 - e^2/r^{N-2}. \tag{6}$$

It can be seen that for $N > 5$ there is no energy minimum. For $N = 4$ the situation is ambiguous because there ceases to exist any characteristic length in the system. This also indicates that no minimum energy scale can exist. It is possible to demonstrate this more rigorously by including special relativistic effects in the energy equation (6). Thus, for $N = 4$, the relativistic energy is (where m_0 is the rest mass of the electron now),

$$E \approx (c^2 h^2/r^2 + m_0^2 c^4)^{\frac{1}{2}} - e^2/r^2, \tag{7}$$

and so as $r \to 0$, $E \to -1/r^2$ and no stable minimum can exist.

On the basis of these arguments it could be claimed that, if we assume the structure of the laws of physics to be independent of the dimension, stable atoms, chemistry and life can only exist in $N < 4$ dimensions. (Note that in two dimensions all energy levels are discrete and there exists a finite energy minimum together with a spectrum extending to infinity, the radius of the first orbit is huge, *ca.* 0.5 cm.) These simple arguments can be confirmed by solving the Schrödinger equation for the N-dimensional hydrogen atom. The dimensionality of the Universe is a reason for the existence of chemistry and therefore, most probably, for chemists also.

† A much stronger, *nonlinear* constraint is required in addition to the Heisenberg uncertainty principle if one is to rule out ground state energies becoming arbitrarily negative. The strongest result is supplied by the nonlinear Sobelev inequality (Lieb 1976). This supplies the required bound on the ground state energy and shows that matter is indeed stable in quantum theory.

4. Wave equations

Many authors have drawn attention to the fact that the properties of wave equations are very strongly dependent upon the spatial dimension (Ehrenfest 1917; Poincaré 1917; Hadamard 1923). Three-dimensional worlds appear to possess a unique combination of properties which enable information-processing and signal transmission to occur via electromagnetic wave phenomena. Since our Universe appears governed by the propagation of classical and quantum waves it is interesting to elucidate the nature of this connection with dimensionality.

Let us recall, as simple examples, the solutions to the simple classical wave equation in one, two and three dimensions.

One dimension:

$$\frac{1}{c^2}\frac{\partial^2 u}{\partial t^2} = \frac{\partial^2 u}{\partial x^2}, \tag{8}$$

where c is the signal propagation speed,
with initial conditions set at $t = 0$ as

$$\left.\begin{array}{l} u(x, 0) = f(x), \\ \dfrac{\partial u}{\partial t}(x, 0) = g(x). \end{array}\right\} \tag{9}$$

This has the solution of D'Alembert,

$$u(x, t) = \tfrac{1}{2}[f(x+ct) + f(x-ct)] + \frac{1}{2c}\int_{x-ct}^{x+ct} g(y)\,\mathrm{d}y. \tag{10}$$

Two dimensions:

$$\frac{1}{c^2}\frac{\partial^2 u}{\partial t^2} = \frac{\partial^2 u}{\partial x^2} + \frac{\partial^2 u}{\partial y^2}, \tag{11}$$

with initial conditions at $t = 0$ of

$$\left.\begin{array}{l} u(x, y, 0) = f(x, y), \\ \dfrac{\partial u}{\partial t}(x, y, 0) = g(x, y). \end{array}\right\} \tag{12}$$

This has the solution of Poisson,

$$u(x, y, t) = \frac{1}{2\pi c}\frac{\partial}{\partial t}\iint_{\rho \leqslant ct} \frac{f(\xi, \eta)}{(c^2 t^2 - \rho^2)^{\frac{1}{2}}}\,\mathrm{d}\xi\,\mathrm{d}\eta + \frac{1}{2\pi c}\iint_{\rho \leqslant ct} \frac{g(\xi, \eta)}{(c^2 t^2 - \rho^2)^{\frac{1}{2}}}\,\mathrm{d}\xi\,\mathrm{d}\eta. \tag{13}$$
$$\equiv \rho^2 = (\xi - x)^2 + (\eta - y)^2.$$

Three dimensions:

$$\frac{1}{c^2}\frac{\partial^2 u}{\partial t^2} = \frac{\partial^2 u}{\partial x^2} + \frac{\partial^2 u}{\partial y^2} + \frac{\partial^2 u}{\partial z^2}, \tag{14}$$

with initial conditions at $t = 0$ as

$$\left.\begin{array}{l} u(x, y, z, 0) = f(x, y, z), \\ \dfrac{\partial u}{\partial t}(x, y, z, 0) = g(x, y, z). \end{array}\right\} \tag{15}$$

This has the solution of Kirchoff,

$$u(x, y, z, t) = \frac{1}{4\pi c^2}\frac{\partial}{\partial t}\left[\frac{1}{t}\iint_{r=ct} f(\xi, \eta, \zeta)\,\mathrm{d}S\right] + \frac{1}{4\pi c^2 t}\iint_{r=ct} g(\xi, \eta, \zeta)\,\mathrm{d}S, \tag{16}$$

where $r^2 = (\xi - x)^2 + (\eta - y)^2 + (\zeta - z)^2$ and $\mathrm{d}S$ is the surface element with respect to (ξ, η, ζ) on the sphere $r = ct$ centred on $(x, y, z) = (0, 0, 0)$.

From these three solutions, (8)–(16), something remarkable emerges. We see that in the one- and two-dimensional cases, the domain of dependence that determines the solution $u(x,t)$ at point (x,t) is given by the closed interval $[x-ct, x+ct]$ and the disk (interior plus boundary) $r \leq ct$, respectively. Therefore, in both cases the signals may propagate at any speed less than or equal to c. In complete contrast, the three-dimensional solution has a domain of dependence consisting only of the *surface* of the sphere of radius ct. All three-dimensional wave phenomena travel *only* at the wave velocity c. We are ignoring the effect of dispersion here.

What this means in practice is that in two-dimensional spaces signals emitted at different times can be received simultaneously: signal reverberation occurs. It is impossible to transmit *sharply* defined signals in two dimensions, for example, by waves on a liquid surface. Now it is known that the transmission of wave impulses in a reverberation-free fashion is impossible in spaces with an *even* number of dimensions (Hadamard 1923). The favourable odd dimensional cases are said to obey Huygen's Principle (Courant & Hilbert 1962). This situation has led many to suppose that life could only exist in an odd dimensional world because living organisms require high fidelity information transmission at a neurological or mechanical level (Poincaré 1917; Whitrow 1955).

Interestingly, one can narrow down the number of reasonable odd dimensional spaces even more dramatically by appealing to the need for wave signals to propagate *without distortion*. Three-dimensional worlds allow spherical waves of the form

$$u(x_1, x_2, x_3; t) = h(r) f(r - ct), \tag{17}$$

with
$$r^2 = \sum_{i=1}^{N} x_i^2, \tag{18}$$

to propagate in *distortionless* fashion to large distances; but this is no longer the case for $N > 3$. For example, in seven dimensions Ehrenfest shows that a solution to the wave equation is,

$$u(x_1, \ldots, x_7; t) = (A/r^5) f(t - r/c) + (B/r^4) f'(t - r/c) + (D/r^3) f''(t - r/c), \tag{19}$$

where A, B and D are constants. Thus, at time t there is no reverberation; only signals which were emitted at the time $(t - r/c)$ are received. However, these signals are now strongly distorted because at large r the terms in f'' and f' determine the form of $u(x,t)$.

Only three-dimensional worlds appear to possess the 'nice' properties necessary for the transmission of high fidelity signals because of the simultaneous realization of reverberationless and distortionless propagation.

5. Fundamental units

Our Universe appears to possess a collection of fundamental or 'natural' units of mass, length and time that can be constructed from the physical constants G, h and c, (see Barrow 1983). A dimensionless constant can only be constructed if the electron charge e is also admitted and then we obtain the dimensionless quantity e^2/hc first found by Sommerfeld. In a world with N dimensions the units of h and c remain ML^2T^{-1} and LT^{-1} but the law of gravitation changes in accord with (1) and hence the units of G are $M^{-1}L^NT^{-2}$. Likewise, Gauss's theorem relates e to the spatial dimension and the units of e^2 are ML^NT^{-2}. Thus in N dimensions the dimensionless constant of Nature is proportional to

$$h^{2-N} e^{N-1} G^{\frac{1}{2}(3-N)} c^{N-4}. \tag{20}$$

It is interesting to notice that for $n = 1, 2, 3, 4$, the constants of electromagnetism, quantum theory, gravity and relativity are absent respectively. Only for $N > 4$ are they all included in a single dimensionless unit. A physical explanation of this result would be very enlightening, especially in view of the role that gauge theories with $N > 3$ may place in effecting a unification of all the fundamental forces (Salam & Strathdee 1982; Cremmer & Julia 1979; de Wit & Nicolai 1982; Cremmer et al. 1978; Witten 1981, 1982).

6. Mathematics

So far, we have displayed a number of special features of physics in three dimensions under the assumption that the form of the underlying differential equations do not change with dimension. One might suspect the form of the laws of physics to be special in three dimensions if only because they have been constructed solely from experience in three dimensions. If we could live in a world of seven dimensions perhaps we would end up formulating its laws in forms that made seven dimensions look special. One can test the strength of such an objection to some extent by examining whether or not 3 and $(3+1)$ dimensions lead to special results in pure mathematics where the bias of the physical world should not enter. Remarkably, it does appear that low-dimensional groups and manifolds do have anomalous properties. Many general theorems remain unproven or are untrue only in the case of $N = 3$; a notable example is Poincaré's theorem that a smooth N dimensional manifold with homotopy type S^N is homeomorphic to S^N. This theorem is true if $N \neq 3$ and the homeomorphism can be replaced by a diffeomorphism if $N = 1, 2, 5$ or 6 (the $N = 4$ case is open). Other examples of this ilk are the problem of Schoenflies and the annulus problem; each has unusual features when $N = 3$. In addition, the low-dimensional groups possess many unexpected features because of the 'accidental' isomorphisms that arise between small groups. The twistor programme (see, for example, Penrose 1977), takes advantage of some of these features unique to $(3+1)$-dimensional space–times.

There is one simple geometrical property unique to three dimensions that plays an important role in physics: universes with three spatial dimensions possess a unique correspondence between rotational and translational degrees of freedom. Both are defined by only three components. In geometrical terms this dualism is reflected by the fact that the number of coordinate axes, n, is only equal to the number of planes through pairs of axes, $\frac{1}{2}n(n-1)$, when $n = 0$ or 3. These features are exploited in physics by the Maxwell field. In an $(n+1)$-dimensional space–time, electric, E, and magnetic, B, vectors can be derived from an $(n+1)$ dimensional potential A_i. The field B is derived from $\frac{1}{2}n(n-1)$ components of curl A_i while the E field derives from the n components of $\partial A_i/\partial t$. Alternatively we might say that in order to represent an antisymmetric second rank tensor as a vector, the $\frac{1}{2}n(n-1)$ independent components of the tensor must equal the spatial dimension, n. So the existence of axial vector representations of quantities like the magnetic vector B and the structure of electromagnetic fields is a consequence of the tri-dimensional nature of space.

There also exists an interesting property of Riemannian spaces that has physical relevance: in an $(n+1)$-dimensional manifold the number of independent components of the Weyl tensor if zero for $N \leq 2$, and so all the 1, 2 and 3 dimensional space–times will be conformally related and they will not contain gravitational waves. The non-trivial conformal structure for $N = 3$ leads to the properties of general relativity.

DIMENSIONALITY

As a final example where the mathematical consequences of dimensionality spill over into areas of physics we should mention the theory of dynamical systems, or ordinary differential equations,

$$\dot{x} = F(x); \quad x = (x_1, ..., x_N). \tag{21}$$

The solution of the system of equations (21) corresponds to a trajectory in an N-dimensional phase space. In two dimensions the qualitative behaviour of the possible trajectories is completely classified. As trajectories in two dimensions cannot cross without intersecting, the possible stable asymptotic behaviours are simple: after large times trajectories either approach a stable focus (stationary solution) or a limit cycle (periodic solution). However, when $N \geqslant 3$, trajectories can behave in a far more exotic fashion. They are now able to cross and develop complicated knotted configurations without intersecting. All the possible behaviours as $t \to \infty$ are not known for $N \geqslant 3$. When $N \geqslant 3$ it has been shown (Ruelle & Takens 1971; Plykin 1974; Newhouse *et al.* 1978) that the *generic* behaviour of trajectories is to approach a *strange attractor*. This is a compact region containing no foci or limit cycles and where all neighbouring trajectories diverge from each other exponentially in time whether followed forwards or backwards along trajectories, so there is sensitive dependence upon initial conditions. Whereas the simple attractors in the one- and two-dimensional phase spaces have dimension one (foci) or two (limit cycles) strange attractors have a *non-integral dimension*. This is manifested by the strange attractor possessing structure on all length scales. When magnified, any portion of the attractor in phase space is revealed to be just as detailed as was its large scale appearance. So far when we have mentioned 'dimension' in §§ 1–5 we have been referring to *topological* dimension but there exist other concepts of dimension that are more useful in practice. In the case of the strange attractor it has a non-integral fractal or Hausdorff dimension (Hausdorff 1918; Mandelbrot 1977; Russell *et al.* 1980). This concept of dimension gives a measure of the amount of information necessary to specify the structure of the attractor. Suppose we wish to cover a line with a minimum number of line segments of length ϵ, $n(\epsilon)$. We clearly need $n(\epsilon)$ of order ϵ^{-1}. Likewise to cover an area or a volume by elemental squares or cubes of side ϵ we require a minimum $n(\epsilon)$ of order ϵ^{-2} or ϵ^{-3} in each case. To cover a d-dimensional surface we require $n(\epsilon)$ of order ϵ^{-d}. In general, we can define a *fractal* dimension as

$$d = \lim_{\epsilon \to 0} \ln(n(\epsilon))/\ln(\epsilon^{-1}). \tag{22}$$

Strange attractors have non-integral values of $d < N$. This situation arises because strange attractors are not manifolds, but the product of a manifold and a fragmented set called a Cantor set. A simple example of a Cantor set can be generated from the unit line segment [0, 1] as follows: delete the middle third of the line segment and then the middle third from all the resulting segments and so on *ad infinitun*. In table 1 we list, L, the total length of the line segment surviving from the rth stage of the deletion process, the number of pieces, n, and the length of each piece.

After an infinite number of these operations the set that remains is a Cantor set. It clearly has zero measure because the total length of the removed pieces is the infinite geometric progression

$$\sum_{r=1}^{\infty} (\tfrac{2}{3})^r = 1. \tag{23}$$

The Cantor set is nowhere dense and cannot be represented by a set of isolated points. If we choose $\epsilon = 3^{-r}$ and $n(\epsilon) = 2^r$ in (22) then we see its dimension, d, is non-integral

$$d = \lim_{r \to \infty} \ln(2^r)/\ln(3^r) = 0.6294.... \tag{24}$$

The value of d tells us how much information is necessary to specify the location of the set to within a given accuracy.

It appears that the early evolution of the Universe, close to the initial singularity, may have possessed these exotic features and displayed chaotic structure on all length scales. In effect, general relativistic space-times can evolve as though they possess a non-integral value of d. This is the case in the Mixmaster universe (Barrow 1981 a, 1982; Chernoff & Barrow 1983) where an infinite number of space-time oscillations occur in any open time interval about the initial singularity at $t = 0$ and there is structure on all scales.

Table 1. The construction of a Cantor set from the interval $[0, 1]$

line decomposition	step	number of pieces, $n(\epsilon)$	length of each piece, ϵ	total length, L
⊢──────────⊣	0	1	1	1
⊢───⊣ ⊢───⊣	1	2	$\frac{1}{3}$	$\frac{2}{3}$
⊔ ⊔ ⊔ ⊔	2	4	$\frac{1}{9}$	$\frac{4}{9}$
	r	2^r	3^{-r}	$(\frac{2}{3})^r$

7. Is $N > 3$?

The idea that the Universe really does possess more than three spatial dimensions has a distinguished history (Kaluza 1921; Klein 1926; Einstein & Bergmann 1938). These authors sought to associate an extra spatial dimension with the existence of electromagnetism. Under a particular symmetry assumption Einstein's equations in $(4+1)$ dimensions look like Maxwell's equations in $(3+1)$ dimensions together with an additional scalar field. Very roughly speaking one imagines uncharged particles as moving only in the $(3+1)$-dimensional subspace but charged particles move through $(4+1)$ dimensions. Their direction of motion determines the sign of their charge.

Supersymmetric gauge theories have rekindled interest in higher dimensional gauge theories that reduce to the $N = 3$ theory by a particular process of dimensional reduction. A topical example is a $(9+1)$-dimensional supergravity theory advocated by Scherk & Schwarz (1974). By analogy with the original Kaluza-Klein theories we would associate $(3+1)$ of these dimensions with our familiar space-time structure whose curvature is linked to gravitational fields and the other additional dimensions correspond to those of a set of internal symmetries. We perceive them as electromagnetic, weak and strong charges. These extra dimensions are compactified to dimensions

$$L \approx \alpha_*^{-\frac{1}{2}} L_P \qquad (25)$$

where $L_P = (Gh/c^3)^{\frac{1}{2}} \approx 10^{-33}$ cm is the Planck length and $\alpha_* = 10^{-1} - 10^{-2}$ is the gauge coupling at the grand unification energy. Thus, according to such theories the Universe will be fully N dimensional (with $N > 3$) when the big bang is hotter than $ca.\ 10^{17}$ GeV, but all except three spatial dimensions will become compactified to microscopic extent when it cools below this temperature after $ca.\ 10^{-40}$ s. One source of interest in this scenario has been to explore whether there exist $(N+1)$-dimensional cosmological models in which such an evolution naturally occurs with three dimensions expanding while others contract. (Chodos & Detweiler 1980; Freund 1982, 1983). These authors show that one can find anisotropic cosmological solutions to $(N+1)$-dimensional general relativity in which 3 spatial dimensions expand at equal rates while the remaining $(N-3)$ spatial dimensions contract. For example, Chodos & Detweiler take a simple metric of Kasner type in $(N+1)$-dimensional space-time (see also Appelquist et al. 1983),

$$ds^2 = dt^2 - \sum_{i=1}^{N} a_i^2(t) dx_i^2, \tag{26}$$

then the vacuum Einstein field equation have the following solution for all N

$$a_i(t) = t^{p_i}; \quad \sum_{i=1}^{N} p_i = \sum_{i=1}^{N} p_i^2 = 1, \tag{27}$$

so at least one of the expansion rates p_i must be negative. If we require $p_1 = p_2 = p_3 \equiv p_+ > 0$ and $p_4 = p_5 = \ldots = p_N \equiv p_- < 0$ then (27) gives, in general,

$$p_+ = [\sqrt{3} + (N-3)^{\frac{1}{2}}(N-1)^{\frac{1}{2}}]/N\sqrt{3}, \tag{28}$$

$$p_- = [(N-3) - \sqrt{3}(N-3)^{\frac{1}{2}}(N-1)^{\frac{1}{2}}]/N(N-3). \tag{29}$$

For $N = 4$ we have $p_+ = -p_- = 0.5$. In the 1-2-3 dimensions this model expands at the same rate as the radiation-dominated Friedman universe. However, it is straightforward to show (Barrow, unpublished work) that models of the type (27) with monotonically increasing and decreasing axes are unstable because they possess isotropic spatial N-curvature. In general, the $(N+1)$-dimensional cosmological models possess a sequence of Mixmaster oscillations near $t \approx 0$ during which the directions of expansion and contraction are permuted in a quasi-random fashion by the anisotropic curvature (Barrow 1982). No single dimension will collapse monotonically as the Universe expands in overall volume $((a_1 a_2 \ldots a_N) \propto t$ according to (27) for all $N)$ and we will not in general create the situation hypothesized by Chodos, Detweiler and Freund. If dimensional compactification occurs it must have a much subtler origin.

8. The anthropic principle

Even if cosmological models like (27) were stable they would not offer a convincing explanation for the observed $(3+1)$ dimensions. After all, there is no reason why only three dimensions should be left expanding. Since other contributors (Rees, Press, Carter, this volume) are discussing some aspects of the anthropic principle (see Barrow & Tipler 1983) it is interesting to note that Whitrow (1955) first suggested an anthropic 'explanation' for why we observe space to possess three dimensions. Perhaps out of an ensemble of all possible universes of all possible dimensionalities observers can only exist in those with three space dimensions? One approach to providing circumstantial evidence in favour of the anthropic principle has been to show that life-supporting or sustaining aspects of the physical world are very sensitive to slight changes in the values of the fundamental constants of Nature (Dicke 1957, 1961; Carter 1974; Carr & Rees 1979; Barrow 1981b). It is obvious from our discussion above that the consequences of the equations of physics are very sensitive to their dimension because they are differential equations, but when it comes to making small changes in the values of fundamental constants like e or G one is on much shakier ground. Although a small change in either of these quantities might so alter the rate of cosmological or stellar evolution that life could not evolve, how does one know that compensatory changes in other constants might not recreate a favourable set of solutions? Suppose, for simplicity, we treat the laws of physics as a set of n ordinary differential equations that contain a set of constant parameters λ_i which we identify with the constants of physics

$$\dot{x} = F(x; \lambda_i); \quad x \in (x_1 \ldots x_n). \tag{30}$$

The structure of the physical world is represented by the solutions of this system, say x^* for the particular realization, λ_i^* of constants that we observe. Is the solution x^* stable against small changes in the parameter set λ_i^*? This is the type of question that the Ruelle–Takens theory described in § 6 is designed to answer for generic systems of equations of the form (30). It tells us that for $n \geqslant 3$ (which will certainly be the case for our model of the laws of physics) the solution x^* will become unstable to changes in λ_i away from λ_i^* past some critical value. If the original attractor at x^* was a simple non-chaotic one with integral Hausdorff dimension then our set of laws and constants are very special in λ_i space but if the original attractor was strange then there should be many other similar sets in the λ_i parameter space. Whether these attractors have anything to do with the necessary and sufficient conditions for observers is an interesting question.

The author would like to thank Dr R. Fenn for helpful discussions.

References

Appelquist, T., Chodos, A. & Myers, E. 1983 *Physics Lett.* B **127**, 51.
Barrow, J. D. 1981a *Phys. Rev. Lett.* **46**, 963.
Barrow, J. D. 1981b *Q. Jl R. astr. Soc.* **22**, 388.
Barrow, J. D. 1982 *Physics Rep.* **85**, 1.
Barrow, J. D. 1983 *Q. Jl R. astr. Soc.* **24**, 24.
Barrow, J. D. & Tipler, F. J. 1983 *The anthropic cosmological principle*. Oxford: Oxford University Press. (In the press.)
Brouwer, L. E. J. 1911 *Math. Annln* **70**, 161.
Brouwer, L. E. J. 1913 *J. Math.* **142**, 146.
Carr, B. J. & Rees, M. J. 1979 *Nature, Lond.* **278**, 605.
Carter, B. 1974 In *Confrontation of cosmological theories with observation* (ed. M. S. Longair). Dordrecht: Reidel.
Chernoff, D. F. & Barrow, J. D. 1983 *Phys. Rev. Lett.* **50**, 134.
Chodos, A. & Detweiler, S. 1980 *Phys. Rev.* D **21**, 2167.
Courant, R. & Hilbert, D. 1962 *Methods of mathematical physics*. New York: Interscience.
Cremmer, E. & Julia, B. 1979 *Nucl. Phys.* B **159**, 141.
Cremmer, E., Julia, B. & Scherk, J. 1978 *Physics Lett.* B **76**, 409.
de Wit, B. & Nicolai, H. 1982 *Physics Lett.* B **108**, 285.
Dicke, R. 1957 *Rev. mod. Phys.* **29**, 375.
Dicke, R. 1961 *Nature, Lond.* **192**, 440.
Ehrenfest, P. 1917 *Proc. Amst. Acad.* **20**, 200.
Einstein, A. & Bergmann, P. 1938 *Ann. Math.* **39**, 683.
Freund, P. G. O. 1982 *Nucl. Phys.* B **209**, 146.
Freund, P. G. O. 1983 *Physics Lett.* B **120**, 335.
Hadamard, J. 1923 *Lectures on Cauchy's problem in linear partial differential equations*. New Haven: Yale University Press.
Handyside, J. 1929 (transl.) *Kant's inaugural dissertation and early writings on space*, pp. 11–15. Chicago: Open Court.
Hausdorff, F. 1918 *Math. Annln* **79**, 157.
Hurewicz, W. & Wallman, H. 1941 *Dimension theory*. New Jersey: Princeton University Press.
Kaluza, T. 1921 *Sber. preuss. Akad Wiss Phys. Math. Kl.* 966.
Klein, O. 1926 *Z. Phys.* **37**, 895.
Lieb, E. 1976 *Rev. mod. Phys.* **48**, 553.
Mandelbrot, B. 1977 *Fractals: form, chance and dimension*. San Francisco: Freeman.
Menger, K. 1928 *Dimensions theorie*. Leipzig: Teubner.
Neugabauer, O. 1975 *A history of ancient mathematical astronomy*, (part 2), p. 848. New York: Springer.
Newhouse, S., Ruelle, D., Takens, F. 1978 *Communs math. Phys.* **64**, 35.
Penrose, R. 1977 *Rep. math. Phys* **12**, 65.
Plykin, R. 1974 *Sb. Math.* **23**, 333.
Poincaré, H. 1917 *Dernières pensées*. Paris: Flammarion.
Ptolemy, C. 1907 *Opera* II, p. 265 (ed. J. L. Heiberg). Leipzig: Teubner.
Ruelle, D. & Takens, F. 1971 *Communs math. Phys.* **20**, 167.
Russell, D. A., Hanson, J. D. & Ott, E. 1980 *Phys. Rev. Lett.* **45**, 1175.
Salam, A. & Strathdee, J. 1982 *Ann. Phys.* **141**, 316.
Sarton, G. 1959 *History of science*, vol. 1, pp. 438–9. New York: Norton.
Scherk, J. & Schwarz, J. H. 1974 *Nucl. Phys.* B **81**, 118.
Sneddon, I. N. & Thornhill, C. K. 1949 *Proc. Camb. phil. Soc.* **45**, 318.
Weyl, H. 1922 *Space, time, matter*, p. 284. New York: Dover.
Whitrow, G. J. 1955 *Br. J. Phil. Sci.* **6**, 13.
Witten, E. 1981 *Nucl. Phys.* B **186**, 412.
Witten, E. 1982 *Nucl. Phys.* B **195**, 481.

Phil. Trans. R. Soc. Lond. A **310**, 347–363 (1983)
Printed in Great Britain

The anthropic principle and its implications for biological evolution

BY B. CARTER, F.R.S.

*Groupe d'Astrophysique Relativiste, Obervatoire de Paris – Meudon,
5 Place Jules Janssen, 92 Meudon, France*

In the form in which it was originally expounded, the *anthropic principle* was presented as a warning to astrophysical and cosmological theorists of the risk of error in the interpretation of astronomical and cosmological information unless due account is taken of the biological restraints under which the information was acquired. However, the converse message is also valid: biological theorists also run the risk of error in the interpretation of the evolutionary record unless they take due heed of the astrophysical restraints under which evolution took place. After an introductory discussion of the ordinary ('weak') anthropic principle and of its more contestable ('strong') analogue, a new application of the former to the problem of the evolution of terrestrial life is presented. It is shown that the evidence suggests that the evolutionary chain included at least one but probably not more than two links that were highly improbable (*a priori*) in the available time interval.

1. INTRODUCTION

A key event in the birth of modern science at the time of the renaissance was the Copernican revolution that transformed our understanding of our planetary system by de-throning the Earth from its central role in favour of the Sun. This was the beginning of a consistent effort by scientifically minded thinkers to break away from the anthropocentric prejudices that had dominated the mediaeval outlook. At the outset, this trend was entirely justified by the goal of scientific objectivity, but it soon came to be carried unduly far as people came to the point of advocating the opposite extreme point of view, consisting in the assumption that our own situation in the Universe is not in any way privileged, but is typically representative in a Universe that is entirely homogeneous apart from minor local fluctuations. This extreme antithesis of the anthropocentric outlook was most dangerous as a source of biased thinking when it was adopted subconsciously. However, it became easier to cope with after having been formulated explicitly as the 'perfect cosmological principle' by Bondi & Gold (1948), who used it as a hypothesis in setting up the 'steady state theory'. (After having been developed in more detail by other workers, starting with Hoyle (1949), the steady state idea fell into general disfavour for a number of theoretical and observational reasons, but a more sophisticated, albeit circumscribed, version known as the 'inflationary universe' has recently resuscitated this perennially beguiling concept (Hawking 1982).)

It was in an attempt to draw attention to the need for a more balanced intermediate attitude, between primitive anthropocentrism and its equally unjustifiable antithesis that I came to introduce the term *anthropic principle* (Carter 1974) to express the notion that 'although our situation is not necessarily central it is necessarily privileged to some extent', in so much as special conditions are necessary for our very existence. The practical scientific utility of this principle arises from its almost tautological corollary to the effect that in making general inferences from what we observe in the Universe, we must allow for the fact that our observations are inevitably biased by selection effects arising from the restriction that our situation should satisfy the conditions that are necessary *a priori*, for our existence. The term *self-selection principle* would be an alternative and perhaps more appropriate description for this hardly questionable but easily overlooked precept. (If I had guessed that the term 'anthropic principle' would come to be so widely adopted I would have been more careful in my original choice of words. The imperfection of this now standard terminology is that it conveys the suggestion that the principle applies only to mankind. However, although this is indeed the case as far as we can apply it ourselves, it remains true that the same self-selection principle would be applicable by any extraterrestrial civilization that may exist.)

In a typical application of the anthropic (self-selection) principle, one is engaged in a scientific discrimination process of the usual kind in which one wishes to compare the plausibility of a set of alternative hypotheses, $H(T_i)$, say, to the effect that respectively one or other of a corresponding set of theories $T_1, T_2, \ldots$ is valid for some particular application in the light of some observational or experimental evidence, E, say. Such a situation can be analysed in a traditional Bayesian framework by attributing *a priori* and *a posteriori* plausibility values (i.e. formal probability measures), denoted by p_E and p_S, say, to each hypothesis respectively before and after the evidence E is taken into account, so that for any particular result X one has

$$p_E(X) = p_S(X/E), \tag{1.1}$$

the standard symbol / indicating conditionality. According to the usual Bayesian formula, the relative plausibility of any two theories A and B, say, is modified by a factor equal to the ratio of the corresponding conditional *a priori* probabilities $p_S(E/A)$ and $p_S(E/B)$ for the occurrence of the result E in the theories, i.e.

$$\frac{p_E(A)}{p_E(B)} = \frac{p_S(E/A)}{p_S(E/B)} \frac{p_S(A)}{p_S(B)}. \tag{1.2}$$

Now in the practical application of this formula it is important to bear in mind that the result will not be valid unless all relevant effects of experimental bias and observational selection have been taken into account in the interpretation of the probabilities on the right hand side. In other words one must be careful to distinguish between the appropriately renormalized *a priori* probabilities that we have denoted by p_S (S for selected or subjective) which are effective here, and the raw *ab initio* probabilities, which could conveniently be denoted by p_O (O for original or objective), that one might have derived directly from the purely abstract theory without taking account of the practical details of its concrete application. The relation between the *a priori* (selected) and *ab initio* (original) probabilities of a result X is expressible, analogously to (1.1) as

$$p_S(X) = p_O(X/S), \tag{1.3}$$

where S denotes the totality of all the selection conditions that are implied by the hypothesis of application of the theory to a concrete experimental or observational situation, but which are not necessarily included in the abstract theory on which the calculation of the *ab initio* probabilities is based. The distinction on which I have been rather laboriously insisting is entirely familiar to all working empirical scientists (even though it is easily forgotten by pure theoreticians who prefer to work exclusively at the *ab initio* level rather than at the practically relevant *a priori* level). The only new element brought in by the anthropic principle is the reminder that the set of subjective selection conditions, S, should include not only the usual allowance for the limitations of our (artificial) measuring instruments but also allowance for our own limitations as living organisms.

In order to illustrate the rather abstract considerations that have just been summarized, let us start with a quite trivial example in which the anthropic aspect as such is not involved. In an investigation of loss of wheat from a barn, let us suppose that we wish to choose between theory A, which attributes a major share of responsibility to mice, and theory B, which attributes the responsibility almost exclusively to rats, and that these theories seem equally plausible *a priori*. Imagine that we set out to acquire experimental evidence E by setting a trap, and that the first animal we catch is a mouse. Proceeding as pure theoreticians at the *ab initio* level, we would obviously assume that the likelihood of a mouse turning up first was high in the first theory, $p_0(E/A) \approx 1$, but that it would be much lower in the second theory, $p_0(E/B) \ll 1$. If we incautiously used the corresponding ratio, $p_0(E/A)/p_0(E/B) \gg 1$ in the Bayesian formula for calculating the *a posteriori* plausibility ratio from the *a priori* plausibility ratio, we would end by concluding that the mouse theory, A, was the most likely. However, an experienced empirical investigator would certainly consider the possibility of limitations in his equipment before coming to a final conclusion. It might be that all we had at our disposal was an ordinary mouse trap, so that the effective 'capture cross section' for a rat would be negligible. Taking the consequent selection conditions into account we would obtain not only $p_S(E/A) \approx 1$, but also (unless mice were very rare indeed) $p_S(E/B) \approx 1$. Substitution of the resulting ratio $p_S(E/A)/p_S(E/B) \approx 1$ in the formula (1.2) would leave the *a posteriori* plausibility ratio unchanged from its *a priori* value, i.e. our experiment would have failed to discriminate.

As a second illustration, I shall now describe an equally simple but non-trivial case, which constitutes what is in fact the classic example of an argument based on the anthropic principle. This example, which is logically analogous to the preceding one, is concerned with the important question (which will be dealt with again in a more conclusive manner later on) of discriminating between hypothesis A to the effect that the development of life is of common occurrence on 'habitable' planets, comparable with our own, and hypothesis B to the effect that, on the contrary, life is very rare, even in geophysically favourable conditions. The evidence, E, consists of the fact that on the only obviously 'habitable' planet we have yet been able to observe, namely our own, life does indeed exist. If future astronomical progress should one day enable us to observe a second example of occurrence of life on a randomly chosen 'habitable planet' belonging to a not too distant star in our Galaxy, the corresponding *ab initio* probability ratio, $(p_0(E/A)/(p_0(E/B)) \gg 1$, would justify the induction that hypothesis A (that life is common) was the most likely. However, so long as the only example at our disposal is our own, no such inference is permissible, since the anthropic selection principle ensures, as a virtual tautology, that one of the *a priori* conditions, S, that must be satisfied by the first planet available for investigation by us must be the prior occurrence of life, namely our own. Thus as in the previous example we obtain not only $p_S(E/A) = 1$ but also $p_S(B) = 1$, so that our observation has no discriminating power at all, and both alternatives A and B remain equally viable. (We shall show later on that when further evidence and more detailed anthropic selection effects are taken into account it is possible to infer that hypothesis A (that life is common) is actually far less plausible than hypothesis B (that life is rare), at least so long as one has no deep theoretical reason for preferring one hypothesis to the other *a priori*.) This example was discussed at length not long ago by Crick (1981) who at the time was apparently well acquainted with the concept, though not the name, of the anthropic principle. With reference to the commonly made but unjustifiable inference that A is most likely, Crick appropriately commented, 'this argument is false', adding, 'I do not know whether such a line of reasoning has a name, but it might be called the *statistical fallacy*'. A more explicit description is in fact already available, namely *failure to respect the anthropic principle*.

As a third illustration, I would mention the case that first drew my own attention to this matter, which was an equally (albeit, at the time, less obviously) unjustifiable inference by Dirac (1937) in favour of a cosmological hypothesis A to the effect that the strength of gravitational coupling diminishes with time. In this example the role of hypothesis B is taken by the orthodox assumption that ordinary general relativistic theory, with a fixed gravitational coupling strength, is appropriate. Dirac's evidence E was effectively equivalent to the observation of a remarkable coincidence which is that the order of magnitude, *ca.* 10^{58}, of the Hubble time t_H (which can be roughly interpreted as indicating the order of the age of the Universe) obtained from the expansion rate of the Universe, as measured in fundamental Planck units ($G = c = h = 1$), agrees with the $\frac{3}{2}$ power of the gravitational coupling constant specified by the square of the proton mass in the same units, i.e. $m_p^2 \approx 10^{38}$. According to Dirac's hypothesis, A, this gravitational coupling varies so as to preserve this relation with the Hubble time as the latter increases with the age of the Universe, thereby giving the result, $p_O(E/A) \approx 1$, that was desired. Whereas in the orthodox theory B the changing Hubble age can agree with the fixed power of the coupling constant only at one particular epoch so that the *ab initio* likelihood of the occurrence of the coincidence is low, i.e. $p_O(E/B) \ll 1$. However, as in the preceding examples, this does not justify discrimination against B in practice, because there is a selection condition (of an anthropic nature) which ensures that the relevant *a priori* probability is nevertheless of order unity, i.e. $p_S(E/B) \approx 1$ as before. The selection effect in question, which was first pointed out by Dicke (1961) consists in the condition that biological systems based on the same principles as our own can hardly come into existence except when the age of the Universe is of the same order as the well known main-sequence (hydrogen burning) lifetime, $\tau_0 \approx m_p^{-3}$, of a typical ordinary star. (The reason, briefly, is that some stars must already have burned out to provide the medium elements that are essential for our chemical constitution, while the maintenance of a suitable continuous energy supply requires that some stars – including the Sun in our own case – must still be burning.)

2. The strong anthropic principle: a digression

With reference to the analysis that has just been presented of the significance of Dirac's large number coincidence, it is often objected that Dicke's restriction on the cosmological epoch is only applicable to life forms based on principles similar to our own, and need not be valid for quite different life forms whose base of support might be quite different from an ordinary star–planet system. It must certainly be admitted that the difficulty of imagining the technical mechanisms on which such alternative life forms might be based can in no way be used as an argument against their existence (particularly in view of the fact that we do not yet understand all the essential mechanisms in our own case). However, even if we suppose for the sake of argument that such alternative life forms are actually more common than our own, their existence is nevertheless quite irrelevant as far as the preceding line of argument is concerned, because the known fact that we ourselves belong to a star–planet based life system must in any case be taken into account for example by including it (in S) as one of the *a priori* restrictions that should be allowed for in estimating the effective *a priori* probability of obtaining the observed value of the Hubble time. Indeed if the basic theory of stellar structure had been as well understood then as it is today, it would have been possible to use the foregoing considerations to predict the order of magnitude of the cosmological expansion rate that we (as a star–planet based life form) could expect *in advance* of its observational discovery by Hubble (in much the same way as Gamow actually did predict in advance the order of magnitude of the cosmic background radiation). The fact that we observe a value predictable in advance on the basis of the orthodox theory (B) can hardly be used as a justification for adopting an alternative theory (A) as Dirac wished to do.

Another way of reaching the same conclusion is by operating at an earlier level in the succession of inferences. There is no absolute distinction between the facts that are considered as *a priori* restrictions (S) and those that are considered as *a posteriori* evidence (E): all that matters is that they should be taken into account whenever available and relevant. We could perfectly well choose to operate at a more fundamental level, treating the fact that we belong to a star–planet based life system as *a posteriori* evidence (in E) rather than as an *a priori* restriction (in S). If, as devil's advocates, we were also to adopt the assumption that most intelligent observers in the Universe belong to other quite differently based life forms, the observed Hubble time and the concomitant fact that we belong to a star–planet based system would not be predictable in advance: in this representation, unlike the previous one, we would have $p_S(E/B) \ll 1$. However (unless it were claimed that varying gravitational coupling in some way increases the odds in favour of star–planet based life forms) the defenders of Dirac's theory would also have to face the fact that our own case was exceptional, i.e. $p_S(E/A) \ll 1$, again in contrast with the previous representation. A really persistent devil's advocate might still try to wriggle out of this by suggesting that the odds in favour of star–planet based life forms might be higher within the framework of theory A, but (in the absence of explicit ideas about the nature of the alternative life forms) this objection would carry no weight whatsoever, because the prosecution could make the analogous claim in favour of B with equal plausibility. Thus the final comparison factor, $p_S(E/A)/p_S(E/B) \approx 1$, would still be the same as before.

Although (as we have just seen) it is not directly pertinent to the examples that have been considered above, the conceivable (if not explicitly imaginable) existence of radically different life forms is certainly relevant to applications of what I have called the 'strong anthropic principle', which is the analogue, at a more fundamental level of inference, of the ordinary 'weak' anthropic principle that has been the subject of discussion so far. As I originally formulated it (Carter 1974) this 'strong' principle consisted in the remark that our mere existence as intelligent observers imposes restrictions not just on our situation but even on the general properties of the Universe, including the values of the fundamental parameters that are the subject of the present meeting. Although this 'principle' has aroused considerable enthusiasm in certain quarters, it is not something that I would be prepared to defend with the same degree of conviction as is deserved by its 'weak' analogue.

The strong principle is less satisfactory than the weak one for two distinct kinds of reason. First, it is not so evident that it is really applicable even in the conservative form stated above, because on the one hand it is not clear that the unified theories towards which we are progressing (as described in other contributions to this discussion meeting) will ultimately leave over any parameters that are 'fundamental' in the sense of being independently variable in such a manner as to be meaningfully selectable, and even if so, it is, on the other hand, not clear either whether (in view of our ignorance of alternative life forms) these values really would be restricted in the way one might expect naïvely from our own case, on the basis of examples suggested, for example, by Hoyle (1954), Carter (1974, 1976), Carr & Rees (1979), Barrow & Silk (1980) and Nanopoulos (1980). (General discussions have been given by Davies (1980), Demaret & Barbier (1981), Breuer (1982) and Barrow & Tipler (1983); see also the contributions by Barrow and by

Press to the present discussion meeting.) Second, even if it is valid in the above sense, this 'strong' principle cannot be used to make actual predictions except by selecting a 'cognizable' subset within the framework of a hypothetical (and not particularly plausible) ensemble of universes whose existence is commonly assumed to be included implicitly in a less conservative formulation of the 'strong anthropic principle'.

Even the choice of the term 'anthropic' is less judicious in the 'strong' than in the 'weak' case: in retrospect, I regret not having used an expression, such for example as 'the cognition principle', having a more transcendent connotation. The philosopher Gale has recently gone so far as to suggest (Gale 1981) that (in conjunction with the world ensemble hypothesis) this *cognizability* principle might be promoted to the status of a 'reality' principle, but I would like to dissociate myself vigorously from such a proposition, which apparently reflects a widespread misconception among philosophers to the effect that science is concerned with 'reality', whereas actually (unlike philosophy and theology) science is only concerned with 'realism'. The same misconception would appear to be implicit in the doctrine that scientific theories are never verifiable but only falsifiable. This doctrine would of course be justified if one considered that theories should precisely represent universal 'truth', but by such a standard all existing scientific theories are not only falsifiable, but may safely be assumed in advance to be false, which has the implication that (according to the standards imposed by philosophers!) science has so far achieved nothing at all. In practice, however, science is not concerned with underlying truth but more modestly (and, by its own criteria, more successfully) with providing the most simple, coherent and comprehensive possible description of *appearance* (desiderata such as objectivity being biproducts of these requirements). (For example it is for theologians to decide whether or not the devil created the fossil record as a perfect fake 5986 years ago: the answer to this classic question in no way affects the purely scientific problem of understanding what it was that is represented by the – genuine or fake – record.) Scientific theories should not be judged as true or false, but rather should be evaluated as relatively good or bad on the basis of criteria such degree of accuracy, range of applicability, etc. The best theories can predict results in advance, but even partial historical explanations or mere botanical classification of previously known results should not be dismissed as valueless. Applications of the 'strong anthropic principle' should be judged by the standards of this humbler, merely explicative rather than predictive category.

In the remainder of this discussion we shall only be concerned with applications of the ordinary 'weak' anthropic principle whose genuinely predictive power should become apparent, even if it is not already evident from the examples of the preceding section.

3. The remarkable coincidence between the timescale of past biological evolution on Earth and the future life expectancy of the Sun

I now come to the first significant new point that I wish to make in the present discussion, which concerns the relevance of the anthropic principle for the interpretation of the observational evidence pertaining to biological evolution on Earth by the Darwinian selection process. To start with I would like to draw the attention of biologists to an application (Carter 1982) which is very simple, but not quite so obvious as Crick's 'statistical fallacy' (that was described in §1). This application is based on a hitherto neglected numerical coincidence that I personally consider to be much more significant than, for example, the large number coincidence of Dirac (that was also described in §1).

The coincidence to which I am referring is based on the very well known fact (see, for example, Dickerson & Geiss 1976) that the time t_e, say, that has been taken so far by biological evolution on this planet since its formation is given to within a few tens of percent by

$$t_e \approx 0.4 \times 10^{10} \text{ years} \qquad (3.1)$$

and the almost equally well known fact (see, for example, Hoyle 1955) that the 'main sequence' lifetime, τ_0 say, of the Sun, during which the energy output from steady hydrogen burning can maintain favourable conditions for life on Earth, is estimated to be given with not quite comparable precision by

$$\tau_0 \approx 10^{10} \text{ years}. \qquad (3.2)$$

Now the biological processes that have governed the evolution of life up to the present stage of emergence of civilization and the astrophysical processes determining the lifetime of the Sun have nothing directly to do with each other (the slowness of the former arising from the numerical complexity of living systems, whereas the slowness of the latter arises from the weakness of gravitation). Therefore the coincidence of these numbers to within a factor close to two, representing the observation that the Sun is now just about half way through its expected life, does not deserve to be just taken for granted as it seems to have been until now. (Indeed, simply in terms of precision, this coincidence is much more striking than the order of magnitude cosmological coincidences which not unjustifiably caught the attention of Dirac.)

Whereas the principal physical processes governing the lifetime of the Sun are generally believed to be adequately understood, the very complicated mechanisms governing the evolution of living systems cannot yet be analysed, still less predicted, in other than very vague qualitative terms. We certainly do not know enough to predict from first principles whether the expected average time $\bar{t}$ which would be intrinsically most likely for the evolution of a system of 'intelligent observers', in the form of a scientific civilization such as our own, should take much less or much more time than is allowed by the external restraints that limit the duration of favourable conditions. In such a state of ignorance, both of these two alternative possibilities should therefore be retained for consideration as not implausible *a priori*. Only the intermediate borderline case, in which the intrinsically most likely evolution time came out to be of just the same order as the time allowed by external restraints, could be set aside in advance, as being much less plausible *a priori*, and therefore worth considering seriously only if convincing *a posteriori* evidence were obtained against both the other two possibilities.

Now the first of these two possibilities, namely that the intrinsically expected time $\bar{t}$ is very short compared with the externally allowed time τ_0 is indeed rather convincingly excluded by the *a posteriori* evidence that the observed evolution time t_e is not small compared with τ_0, since it is hard to think of any particular reason why our arrival should have been greatly delayed relative to the intrinsically expected time $\bar{t}$. However, provided one avoids making the habitual mistake of overlooking the anthropic principle, it can easily be seen that the observation that t_e is comparable with the upper limit τ_0 is just what would be expected if we adopt the alternative hypothesis that the intrinsically expected time $\bar{t}$ is much longer than τ_0: in this case self-selection ensures that ours must be one of the exceptional cases in which evolution has proceeded much faster than usual; on this basis it is to be expected that t_e should be comparable with τ_0 because there is no particular reason why we should belong to the even more exceptional cases in which evolution proceeds even more rapidly although, with the assumption that the Universe is infinite, such cases must of course exist.

Since this satisfactorily accounts for the observed order of magnitude of t_e, there is no need to have *ad hoc* recourse to the *a priori* less plausible hypothesis that the magnitudes of such unrelated quantities as $\bar{t}$ and τ_0 should just happen to coincide. Although the preceding inductive argument is essentially probabilistic, and hence cannot be absolutely watertight, I consider that it constitutes rather strong evidence for the conclusion that $\bar{t}$ is in fact much larger than τ_0. In particular, this means that there is no justification for the implicit assumption by many science fiction writers that the expectation $\bar{t}$ should be comparable with the observed value t_e. Our present conclusion will no doubt be unpopular in such quarters (which perhaps explains why it has not been pointed out before) because it implies as an obvious corollary – which will be discussed in a rather more quantitative manner in the following sections – that civilizations comparable with our own are likely to be exceedingly rare (even if locations as favourable as our own are of common occurrence in the galaxy, which is by no means evident) so that not much credibility can be attached to the exciting fiction scenarios involving reception of extraterrestrial communications, not to mention visitations.

4. Régimes of chance and necessity in biological evolution

Let us now consider the mechanism of biological evolution in rather more detail. According to the memorable description of Monod (1970) this mechanism is governed by a combination of 'chance and necessity'. The 'chance' here refers to the essentially random mutations in the genetic information passed on by an individual to its immediate descendants, while Monod's 'necessity' refers to the ineluctability with which the Darwinian natural selection process can ultimately impose certain particular kinds of mutation on a species as a whole in suitable circumstances. For many practical purposes, however, Monod's 'necessity' is rather illusory: to start with, there is the limitation (which is particularly relevant to laboratory experiments) that it only applies to very large interbreeding populations; of even greater importance for our present purposes is the consideration that even those changes that really are imposed by natural selection will themselves be functions (and often very sensitive functions) of external ecological conditions. The conditions determining the direction of Darwinian selection, even when it is effective, vary in such a complicated manner owing to the interplay with other species, not to mention geophysical effects, that for the practical purpose of evaluating very long term evolutionary trends they must be treated as being to a very large extent governed by chance. In other words, the relevance of Monod's 'necessity' is effectively confined to an intermediate level, between the stochastically governed régime of microscopic mutation processes and the hypermacroscopic régime of very long term evolutionary trends which is also effectively dominated by random effects.

In order to make these ideas rather more explicit, let us consider some very simple mathematical models that illustrate some of the general qualitative features of evolutionary processes, as they have been understood since the work of pioneers such as Wright (1931). To start with, let us consider a hypothetical biological species in which the hereditary genetic information content carried by each individual is I bits, and in which each separate binary unit is subject to independent random mutations whose rate (per individual per generation) is given by μ/I where the quantity μ so defined is what we shall refer to as the mutation rate factor, and typically can be expected to be very roughly of the order of unity, i.e.

$$\mu \approx 1. \tag{4.1}$$

(It is generally accepted that the genetic information content of existing terrestrial species is carried by nucleic acid chains with a language in which the basic letters are twenty kinds of amino acid that are specified (with some redundancy) by a 6-bit code (in much the same way as is done for the symbols of an ordinary typewriter by the 8-bit ASCII code in modern computer information processing.) If one needed a more precisely realistic model, one would have to allow for the fact that information is duplicated so as to allow sexual interchange, and also that the mutations are not completely independent – the most elementary mutation involving not 1 but 2 bits, corresponding to the four different kinds of nucleotide. However, these complications are unimportant for the crude estimates that will be made here.)

The reason why the rate factors μ can be expected typically to be of the order of unity is that their average value $\bar{\mu}$, as taken over all the I distinct rate factors, is just the *total* rate (per individual per generation) of binary mutations of all kinds occurring in the model. Now this total mutation rate $\bar{\mu}$ cannot be very large compared with unity if the species is to survive because a significant proportion of all mutations can be expected to be lethal. On the other hand it would be disadvantageous for long term adaptability if $\bar{\mu}$ were unnecessarily low. One is therefore led to expect that the internal mechanisms controlling the mutation rate would adjust themselves so as to give values of $\bar{\mu}$ at least roughly of the order of unity, a prediction which has been investigated by a number of workers, and which seems to have been confirmed in many cases. (This reasoning also entails that there should only be rather limited scope for acceleration of evolution by an externally induced increase in mutation rates such as might be induced by exposure to intense radiation.) The total relevant genetic information content I is hard to evaluate experimentally: measurements of total nucleic acid content give reasonably precise upper limits, ranging from the order of 10^9 for bacteria to the order of 10^{10} for mammals, including ourselves (see, for example, Dobzhansky *et al.* 1977), but there is reason to believe that a considerable fraction of this is effectively redundant, so that somewhat lower estimates of I would probably be more appropriate.

Under sufficiently favourable conditions, which include the requirement that the effective interbreeding population number, N say, be large enough, the changes that are favoured by natural selection will be imposed not just 'necessarily' (in Monod's sense) but indeed very rapidly compared with geological timescales. It is easy to see that if a particular kind of mutant is relatively favoured in its breeding by a selection coefficient s (meaning that if no new mutations occurred the mutant fraction, q say, would increase by a factor e^s from one generation to the next) then even if it started from nothing the mutant fraction q would increase so as to reach the order of unity within a timescale T whose order of magnitude, in units of a generation, is given by

$$T \approx s^{-1} \ln(s\mu^{-1}I), \tag{4.2}$$

provided the two following conditions are satisfied. First we must have

$$s \gtrsim \mu I^{-1} \tag{4.3}$$

in order for Darwinian selection to be able to dominate the average mutation rate, and second we require

$$N \gtrsim \mu^{-1}I, \tag{4.4}$$

which is sufficient to ensure that random breeding fluctuations can be neglected. Subject to a more severe restriction on the breeding selection coefficient, s, namely

$$N\mu I^{-1} \gtrsim s \gtrsim N^{-1}, \qquad (4.5)$$

the expression (4.2) will remain applicable even for comparatively small populations in the range

$$N^2 \gtrsim \mu^{-1}I \gtrsim N. \qquad (4.6)$$

If it is assumed that s is not too small and μ not too large compared with unity, their contributions inside the logarithm will be unimportant, and thus the expression (4.2) for what may be thought of as the typical number of generations required for changes by ordinary Darwinian selection effectively reduces simply to $T \approx s^{-1} \ln I$. In view of the fact that sexual interchange within the population makes it possible for many independent kinds of mutation to be undergoing selection in this way all at the same time, one sees that it is in principle possible to make a transition between any two among the 2^I possible information configurations (of which most are of course not viable) within a timescale which depends merely logarithmically on I. Since $\ln I$ is at most in the region of 20 for the human population, we see that even in our own case comparatively modest selection factors, of the order of 1% or so, could bring about major evolutionary changes within a few thousand generations, i.e., in substantially less than 10^5 years. (On the supposition that $\mu^{-1}I$ is in the vicinity of 10^9, the prehistoric human population would probably have been within the range (4.6) most of the time, while the present population has just about reached the range (4.4).)

For extremely small populations the genetic evolution is no longer subject to a régime of 'necessity' but to a régime of 'chance'. Even for quite high values of the Darwinian selection coefficient, s, the initiation of the quasi-deterministic selection process characterized by formula (4.2) will be subject to a stochastic delay, with a characteristic mean timescale τ given in order of magnitude by

$$\tau \approx N^{-1}\mu^{-1}I, \qquad (4.7)$$

which results from the fact that selective breeding cannot begin to operate until the first appearance of mutants of the kind under consideration. Once a few of these mutations have occurred, after an average time given by (4.7), selective breeding will establish the mutant strain comparatively rapidly provided that the condition

$$s\mu^{-1}I \gtrsim N \gtrsim s^{-1} \qquad (4.8)$$

is satisfied. However, if the selective breeding coefficient, s, is too small, another element of 'chance' enters the situation, in so much as random fluctuations (proportional to $N^{\frac{1}{2}}$) in relative numbers of descendants will dominate: thus subject to the necessary conditions

$$\mu^{-1}I \gtrsim N, \quad s^{-1} \gtrsim N, \qquad (4.9)$$

one obtains a 'neutral' régime of what may be described as stochastic genetic drift in which (as described, for example, by Kimura 1979) mutant strains are imposed on the population, or eliminated, at random with a characteristic timescale τ for any particular such kind of mutation given by

$$\tau \approx \mu^{-1}I. \qquad (4.10)$$

(This is the mechanism that is thought to be responsible for the minor differences between analogous proteins in related species, which have been used as a genetic clock for establishing the comparative lengths of the branches of phylogenetic trees.) A similar timescale is obtained for 'neutral' genetic drift in large populations when the selective breeding coefficient s is negligible, which arises when

$$N \gtrsim \mu^{-1}I, \quad s^{-1} \gtrsim \mu^{-1}I; \tag{4.11}$$

but in this case the process is not effectively stochastic, in so much as the mutant fraction q will 'necessarily' increase, slowly and steadily, until it reaches a value of order unity after a time T given by

$$T \approx \mu^{-1}I, \tag{4.12}$$

as a result of simple linear accumulation of mutations in succeeding generations.

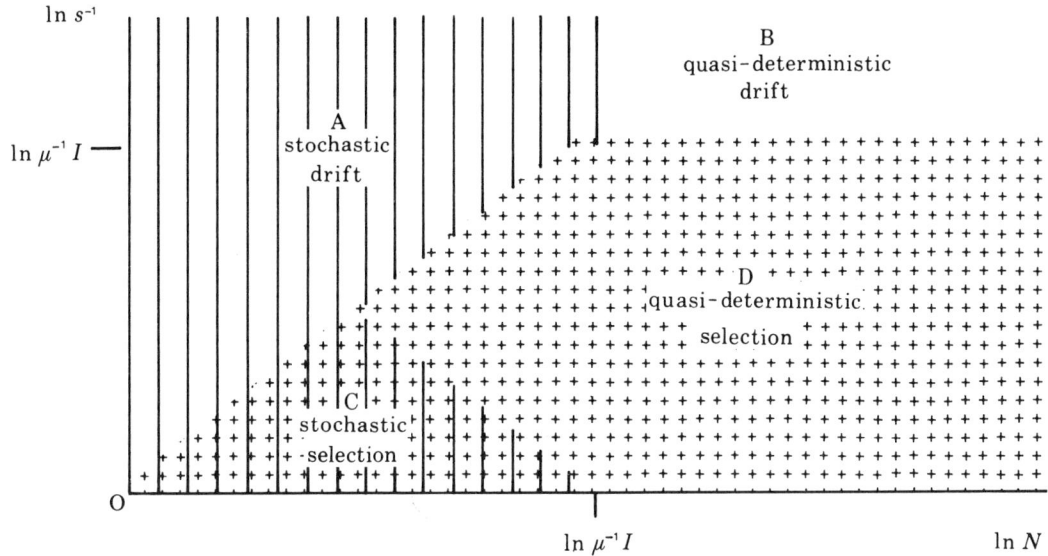

FIGURE 1. The approximate locations of régimes of 'chance' and 'necessity' are shown on a plot of the inverse of the selective breeding coefficient s against the effective breeding population number N with a logarithmic scale normalized with respect to the critical magnitude $\mu^{-1}I$ which represents the inverse of the relevant mutation rate. Darwinian régimes in which natural selection is dominant are marked by crosses, $+++$, and stochastic régimes in which the population is too small for Monod's 'necessity' to apply are marked by vertical shading |||. The four possible combinations are: A the régime of neutral random drift (whose significance was pointed out by Kimura) with characteristic time given by (4.10); B the régime of very slow quasi-deterministic drift on a timescale of the same order, as given by (4.12), C the régime of Darwinian selection in a small population at a stochastically variable rate with characteristic time given by (4.7); finally D the régime of ordinary quasi-deterministic Darwinian selection in a large population, on a potentially short timescale given by (4.2).

The relative positions of the four evolutionary régimes that have just been outlined are plotted on a $\ln N/\ln s$ plane in figure 1.

5. A SIMPLE STOCHASTIC MODEL FOR THE ERRATIC PROCESS OF LONG TERM EVOLUTION

The timescale given by both (4.10) and (4.12) for unselected 'neutral' evolution is very long and indeed in our own mammalian case is comparable with the observed total biological evolution time t_e (which is itself a rather striking coincidence for which I can at present think of no plausible explanation, anthropic or otherwise, in view of the fact that such undirected evolution can hardly have been the primary mechanism in our own emergence even though it must presumably have had a certain subsidiary role). The timescale (4.7) for the stochastic régime of Darwinian selection in small populations will usually be much shorter, but it too is limited by the necessarily low rate, μI^{-1}, of mutations of a given kind per individual per generation.

In contrast with these, the timescale given by (4.2) for ordinary quasi-deterministic Darwinian selection in a large population is virtually independent of the mutation rate. In ordinary time units, as opposed to generations, the corresponding time, t_D say, is expressible very roughly as

$$t_D \approx \tau_g s^{-1} \ln I, \tag{5.1}$$

where τ_g is the average time interval between generations. This time τ_g can be extremely short compared with the observed total evolution time t_e, even for quite moderate values of the selective breeding coefficient s. Thus for bacteria, (with τ_g of the order of a few hours) one can easily obtain $t_D/t_e \approx 10^{-12}$, while even in our own case (with τ_g of the order of tens of years) the corresponding realistic minimum magnitude for this ratio will still have the very low value

$$t_D/t_e \approx 10^{-5}. \tag{5.2}$$

It is thus to be concluded that the time taken by biological evolution on Earth up to our own present stage of advancement has been many tens of thousands of times longer than need have been the case if strong Darwinian selection pressure had at all stages been steadily directed towards the present outcome.

The existence of this delay is consistent with the idea that evolution proceeds largely by fits and starts, in the manner described by Gould & Eldredge (1977). What seems to be implied by the foregoing considerations is that, except for some comparatively brief spurts, either the Darwinian evolution process must have been acting at a negligibly slow rate (as in the much cited example of oysters which seem to have been so well adapted to an effectively unchanging environment that no important mutations have been favourably selected in more than 10^8 years) or else that the Darwinian process has been operating at a comparatively high rate but in an erratic direction (a classic, albeit extreme, example being adaptation from sea to land followed by readaptation to sea). This is to be expected from the consideration that the strongly selected changes (describable in terms of Monod's necessity) that must have occurred at many intermediate stages were not teleologically directed towards our present state or any other long term goal but were directed towards immediate advantages in stochastically changing environmental conditions (the changes in question being largely due to complicated interactions with other species, as well as geophysical effects). Thus, to sum up, one can envisage evolution in terms of comparatively rapid adaptation to niches which are themselves undergoing more or less rapid variation in the space of ecological parameters.

This reasonably well established conclusion brings me to the point at which it is appropriate to introduce further simplifying hypotheses to enable us to set up another rudimentary mathematical model for the purposes of describing evolution up to some particular given state of 'advancement', which we shall later on take to be the formation of a scientific civilization such as our own. This model is based on the supposition that the attainment of a given degree of development of a particular kind (e.g. intellectual, in the application we have in mind) depends on successful passage through a number of intermediate steps involving the acquisition of relevant accessories (e.g. eyes). We further suppose that even in globally favourable circumstances the effective unpredictability of local ecological conditions is such that passage through such a step does not take place automatically but occurs only with a certain probability, λ say, per unit time. The rate λ at which such an intermediate step is likely to occur will depend on the sensitivity of its selective value to variations in ecological conditions. For example, the selective value of even very rudimentary optical radiation detectors would appear to be sufficiently universal to ensure that the appropriate timescale λ^{-1} for the development of eyes is quite short compared with the observed age t_e of the Earth. On the other hand, the acquisition of wings (which in any case is apparently unnecessary for the attainment of our present stage of advancement) would seem to be rather more difficult, having been achieved in only a small number of independent lines of descent, which suggests that such a development should be described in terms of a substantially smaller value of λ (so that λ^{-1} would be of almost comparable magnitude with t_e), due presumably to the fact that rudimentary flying equipment is usually a burden rather than an advantage. As a final example, the appropriate value of λ^{-1} for the biological development of ordinary freely turning wheels (such as are used in considerable numbers in even the simplest man-made machines) would appear (from their observed absence in all terrestrial living organisms as far as I know) to be at least comparable with t_e and probably very much longer. (Such an appearance could, however, be deceptive, as a consequence of the anthropic principle, in the admittedly implausible eventuality that the advent of a – perhaps dangerously competitive – wheel using life forms could diminish the chances of achieving advanced intellectual development.)

Although a realistic model of long term evolution would allow for the fact that the various intermediate steps under consideration will not all be independent, but must in many cases be achieved in a well defined order (e.g. the development of wings would be useless without the prior development of eyes or other long range sensors), it will nevertheless be sufficient for our present very broad and general concerns to ignore such complications and work with a rudimentary model in which the relevant steps are treated as being independent. For the purpose of estimating the likelihood of reaching some given level of development within the astronomically allowed timescale τ_0, the only intermediate steps that need to be explicitly taken into account are those for which the corresponding timescale λ^{-1} is at least a significant fraction of the available time τ_0, since it can be taken for granted that all the others will be achieved with virtual certainty. Thus there will remain only a limited number, n say, of 'critical' steps that are characterized roughly by

$$\lambda^{-1} \gtrsim \tau_0, \tag{5.3}$$

and which must therefore be taken into account. (It follows from the considerations of the preceding section that n must be at least equal to 1.) For the crude qualitative conclusions that we wish to draw here, it will be sufficient, as a final simplification, to treat the n critical steps that are retained for consideration as each having the same 'average' value of λ. In this simple model, there will be an intrinsic probability given by

$$p = (1 - e^{-\lambda t})^n, \tag{5.4}$$

for all n critical steps to have been carried through prior to any given time t. (The application of this model is justifiable provided t is large compared with the characteristic times λ^{-1} of all the non-critical steps that were left out of account, and hence in particular whenever t is at least comparable with τ_0).

Having thus set up the simplest plausible model for the description of long term evolutionary 'progress', let us now consider its implications. To start with we can evaluate the average time $\bar{t}$ by which all the n critical intermediate steps are completed: substitution of (5.4) gives

$$\bar{t} = \int_0^\infty t\,dp = \lambda^{-1} h_n, \qquad (5.5)$$

where

$$h_n = 1 + 1/2 + \ldots + 1/n \qquad (5.6)$$

is the Euler sequence, which for large values of n is given asymptotically by the expression

$$h_n = \ln n + \gamma + O(1/n), \qquad (5.7)$$

where $\gamma = 0.577\ldots$ is Euler's number.

In applying the model to the emergence of a scientific civilization on Earth, we recall the considerations of §3 which suggest that we must have

$$\bar{t} \gg \tau_0, \qquad (5.8)$$

since to have $\bar{t} \approx \tau_0$ would *a priori* be an unlikely coincidence; the remaining alternative, $\bar{t} \ll \tau_0$ is hardly compatible with the observation that the actual time t_e of our emergence satisfies the relation

$$t_e \approx \tau_0. \qquad (5.9)$$

Although no obvious external considerations can affect the intrinsically low probability,

$$p \approx \exp\{-h_n t_e/\bar{t}\} \qquad (5.10)$$

of obtaining $t_e \gg \bar{t}$, on the other hand the upper limit τ_0 on the timescale during which the conditions for applicability of the model are maintained makes it certain, by application of the (ordinary, weak) anthropic principle that we shall have $t_e \lesssim \tau_0$ even if (5.8) is satisfied.

In view of the fact that the distribution dp/dt obtained from (5.4) increases monotonically up to a maximum at a time t_m given by

$$t_m = \lambda^{-1} \ln n \approx \bar{t}, \qquad (5.11)$$

we see that subject to the hypothesis (5.8) the cut-off, $t_e \lesssim \tau_0$, leads to the prediction that the relative probability of obtaining the observed coincidence (5.9) is of order unity. As an immediate corollary of the same observationally consistent hypothesis (5.9) we obtain the prediction that the probability of completing all the n necessary intermediate steps on a randomly chosen planet subject to the same average global environmental conditions as our own will have the very small value

$$p \approx (h_n \tau_0/\bar{t})^n, \qquad (5.12)$$

since for any reasonable value of n, the order of magnitude of the factor h_n cannot be very large compared with unity. Quite moderate lower bounds $n \gtrsim 10$ and $\bar{t}/\tau_0 \gtrsim 10^2$ give $p \lesssim 10^{-20}$, which is more than sufficient to ensure that our stage of advancement is unique in the visible Universe. In view of our utter lack of quantitative understanding of the mechanisms that determine the rates λ, it is by no means implausible that the relevant mean time should have a very much large value, for example $\bar{t}/\tau_0 \gtrsim 10^{10}$, which would be sufficient to ensure that our level of advancement is unique in at least our own galaxy even if the number n of necessary critical steps is as low as one.

6. The number of critical steps: a dilemma?

The preceding interpretation of the conclusions of §3 within the framework of our simplified stochastic model does not exhaust the conclusions that can be drawn from the present application of the anthropic principle. We shall now go on to point out some further implications of a less trivial and more interesting nature.

The power law dependence,

$$p \approx \alpha t^n \qquad (6.1)$$

(for constant α) which is obtained from (5.4) in the limit for $t \ll \bar{t}$ (and which would *still* hold for a more sophisticated model taking account of differing rates λ from one step to another and of the necessity that the critical steps be taken in the correct order) implies that with a relative probability close to unity the completion of the n critical steps within the allowed time range $0 < t < \tau_0$ will occur near the end of this range to within a fraction of the order of magnitude of $1/n$. Thus due allowance for the anthropic principle not only explains the order of magnitude relation (5.9) but also leads to the much more precise prediction

$$\tau_0 - t_e \approx n^{-1}\tau_0. \qquad (6.2)$$

Now, as we have already remarked, the observational coincidence (5.9) is valid not merely as a crude order of magnitude relation but within a factor close to two, in so much as the standard estimates of the duration of the future main sequence life of the Sun give a value of the same order as the age of the Earth, i.e.

$$\tau_0 - t_e \approx t_e, \qquad (6.3)$$

to within an uncertainty factor hardly exceeding 100 %. Thus provided the broad theoretical framework we have been using is essentially correct (and it is to be emphasized that the reasoning in this section remains valid even when one allows for ordering and unequal probabilities of the critical steps) one sees that in order to fit the prediction (6.2) with the observation (6.3) we must have at most a very small number of critical steps: the values $n = 1$ and $n = 2$ are quite consistent, but values from $n = 3$ onwards become rapidly more difficult to reconcile with the comparatively long period during which terrestrial conditions seem likely to remain favourable. In short, combining this conclusion with that of the previous section, we are led to induce that

$$1 \leqslant n \lesssim 2. \qquad (6.4)$$

My first reaction on arriving at this quite severe upper limit on n was one of surprise, since I had previously been inclined to think that the appropriate value of n (i.e. the number of intermediate steps necessary for evolution to our present stage whose rates λ are low on timescales comparable with the age of the Earth) was likely to be very large. This vague prejudice arose from a certain appearance in the fossil record of a consistent trend in the direction of long term evolution towards our present state (the same appearance that in an extrascientific context has often been used as an argument in favour of the idea of 'divine guidance'). One has the impression that successively more recently evolved categories of animals can be classified in a rising hierarchy of levels of increasing 'advancement', this latter term being defined in terms of acquisition of features apparently necessary for the ultimate emergence of civilization. The transitions between these levels would be candidates for treatment as critical steps in the sense of our simple stochastic model, a typical example being the development of the placenta, for which the corresponding rate λ must be fairly low in view of the fact that it has occurred only on one side of the Wallace line. The more or less steady occurrence of progressive steps such as this every 10^8 years or so could be accounted for in a satisfactory manner in terms of our simple model by taking a value of n of the order of at least several tens. However, the appearance of relatively steady progress up the evolutionary ladder is harder to account for in terms of a model in which the value of n is very small.

I can at present imagine two quite distinct ways of resolving the dilemma that has just been posed. The first is to suppose that the appropriate value of n really is large (as is suggested by the immediately preceding considerations) and that the prediction (6.2) (on which the upper limit in (6.4) was based) is invalid due to an overestimation of the available timescale during which geophysical conditions will remain favourable. The example of the phenomenon of ice ages shows that the terrestrial climate is sensitive to factors that are still not well understood, so that it is difficult to exclude the possibility that our environment was destined (had we not emerged to retard or accelerate the process) to become unfavourable due to some as yet unforeseen overheating or overcooling effect within a timescale, τ_e say, only very slightly greater than the present age, t_e, of the Earth. On this hypothesis, our previous line of reasoning would need to be modified by the substitution of τ_e in place of τ_0 in (6.2).

Instead of resorting to such an *ad hoc* hypothesis, the alternative way to resolve the dilemma would be to adopt the conclusion (6.3) bravely at its face value, and to submit the rather tenuous reasons for doubting it to closer scrutiny. This means accepting that at most one or two of the steps in our evolution (e.g. the original establishment of the genetic code, and the final breakthrough in cerebral development) were genuinely critical in the sense of our stochastic model. The implication is that all the other apparently important and not obviously ineluctable steps (such as the development of the placenta), are either less difficult than one might suppose or else are merely incidental and not as essential as is widely believed. My present inclination is to believe that the latter is the most likely. In other words, my best guess is that the correct conclusion to be drawn from the reasoning that has been presented is that many of the salient developments in our evolution were quite unnecessary, as well as having been intrinsically improbable, in so much as many alternative evolutionary pathways would have been compatible with the ultimate emergence of civilization (e.g. there is no obvious reason why it should not have arisen in egg laying animals).

If this last interpretation is correct, it means that the apparent existence of an evolutionary ladder is to a large extent an illusion: an artefact of our still unduly anthropocentric imaginations, which lead us to jump too easily to the conclusion that merely because we happen to possess some particular attribute it must be essential for 'higher development'. (If we had happened to be born – or for that matter, hatched – with wings, they would no doubt be generally regarded as an indispensable status symbol for any life form aspiring to be described as 'advanced'!) This final remark leads me to recapitulate the general moral of this exposition, which is that one should try to steer a moderate course between the Scylla of excessive anthropocentrism and the Charybdis of unjustifiable neglect of anthropic selection effects.

References

Barrow, F. J. & Tipler, F. 1983 (In preparation.)
Barrow, J. D. & Silk, J. 1980 *Scient. Am.* **242**, 127.
Bondi, H. & Gold, T. 1948 *Mon. Not. R. astr. Soc.* **108**, 252.
Breuer, R. 1982 *Das Anthropische Prinzip*. Wien: Meyster.
Carr, B. J. & Rees, M. J. 1979 *Nature, Lond.* **278**, 605.

Carter, B. 1974 In *I.A.U. Symposium* **63**: confrontation of cosmological theories with observational data (ed. M. S., Longair), p. 291. Dordrecht: Reidel.
Carter, B. 1976 In *Atomic physics and fundamental constants* **5** (ed. J. H. Sanders & A. H. Wapstra), p. 650. New York: Plenum Press.
Carter, B. 1982 *Pour la Science* **52**, 55.
Crick, F. 1981 In *Life itself: its origins and nature*. p. 90. New York: Simon and Schuster.
Davies, P. 1980 *Other worlds*. London: J. M. Dent.
Demaret, J. & Barbier, C. 1981 *Rev. Quest. Sci.* **152**, 461.
Dicke, R. H. 1961 *Nature, Lond.* **192**, 440.
Dickerson, R. E. & Geiss, I. 1976 *Chemistry, matter and the Universe*. Reading, Massachusetts: Benjamin.
Dirac, P. A. M. 1937 *Nature, Lond.* **139**, 323.
Dobzhansky, T., Ayala, F. J., Stebbins, G. L. & Valentine, J. W. 1977 *Evolution*. San Francisco: Freeman.
Gale, G. 1981 *Scient. Am.* **245**, no. 6, 114.
Gould, S. J. & Eldredge, N. 1977 *Paleobiology* **3**, 115.
Hawking, S. W. 1982 *Physics Lett.* B **115**, 295.
Hoyle, F. 1949 *Mon. Not. R. astr. Soc.* **109**, 365.
Hoyle, F. 1954 *Appl. J. suppl.* **1**, 121.
Hoyle, F. 1955 *Frontiers of astronomy*. London: Heinemann.
Kimura, M. 1979 *Proc. natn. Acad. Sci. U.S.A.* **76**, 3440.
Monod, J. 1970 *Le hazard et la necessité*. Paris: Seuil.
Nanopoulos, D. V. 1980 *Physics Lett.* B **91**, 67.
Press, W. H. & Lightman, A. P. 1983 *Phil. Trans. R. Soc. Lond.* A **310**, 323.
Salpeter, E. E. 1957 *Phys. Rev.* **107**, 506.
Wright, S. 1931 *Genetics, Princeton* **16**, 97.

Discussion

W. H. McCrea, F.R.S. (*University of Sussex, Brighton BN1 9QH, U.K.*). In Dr Carter's scheme for the evolution of man on Earth, it appears that some form of catastrophic happening plays an essential role. What would the outcome be were such a happening not forthcoming? I ask because it seemed almost as though evolution had to anticipate a catastrophe.

B. Carter. I prefer to use the more neutral term 'cut-off' rather than 'catastrophe' for the natural astrophysical (or geophysical) time limitations in question. (Even effects like the ice age phenomenon occur on timescales so long compared with those characteristic of technical development of civilization that there would be adequate time to prepare countermeasures of avoidance or protection. The term catastrophe would be more appropriate for something like a man made ecological disaster, which is an eventuality that might well be discussed with reference to the anthropic principle, but which has nothing directly to do with the considerations that I have presented on this occasion.) Although the certainty of a long term astrophysical cut-off (and the conceivable possibility of a geophysical cut-off in a less distant future) played an essential role in the foregoing line of argument, such a (future) cut-off evidently was not essential for our (past) development as such. What would occur in an imaginary universe that started off in the same way as our own, but in which such cut-offs were miraculously suspended, is that advanced life would ultimately become far more common than it seems to be in our own Universe now, at least in so far as its density was measured in civilizations per galaxy. However, although there would be a very few rare exceptions that would come into existence under conditions similar to our own, most of the life systems in such a universe would not produce anything describable as a scientific civilization until after the universe had reached such an advanced state of expansion that the average number of civilizations per unit volume of intergalactic space might actually be lower than in our own Universe now (particularly if the number n of critical steps really is less than three as has been suggested).

/530.8C758>C1